The Urban Book Series

The Urban Book Series is a resource for urban studies and geography research worldwide. It provides a unique and innovative resource for the latest developments in the field, nurturing a comprehensive and encompassing publication venue for urban studies, urban geography, planning and regional development.

The series publishes peer-reviewed volumes related to urbanization, sustainability, urban environments, sustainable urbanism, governance, globalization, urban and sustainable development, spatial and area studies, urban management, transport systems, urban infrastructure, urban dynamics, green cities and urban landscapes. It also invites research which documents urbanization processes and urban dynamics on a national, regional and local level, welcoming case studies, as well as comparative and applied research.

The series will appeal to urbanists, geographers, planners, engineers, architects, policy makers, and to all of those interested in a wide-ranging overview of contemporary urban studies and innovations in the field. It accepts monographs, edited volumes and textbooks.

Indexed by Scopus.

Xiaoxi Li · Ye Zhang

Conserving and Managing Historical Urban Landscape

An Integrated Morphological Approach

Springer

Xiaoxi Li
School of Architecture
Harbin Institute of Technology
Shenzhen, China

Ye Zhang
Department of Architecture
National University of Singapore
Singapore, Singapore

ISSN 2365-757X ISSN 2365-7588 (electronic)
The Urban Book Series
ISBN 978-981-19-4221-1 ISBN 978-981-19-4222-8 (eBook)
https://doi.org/10.1007/978-981-19-4222-8

This Springer imprint is published by the registered company Springer Nature Singapore Pte Ltd.
The registered company address is: 152 Beach Road, #21-01/04 Gateway East, Singapore 189721, Singapore

Foreword

Signs of the social, symbolic and functional disintegration of the urban fabric multiply around the world…[1]

Manuel Castells

In a period of rapid urbanization and fast transformation processes, the conservation, restoration and rehabilitation of the historical urban landscape became an important task to overcome the lack of shared meaning and cultural communication that Castells complains. This task is not limited to the restoration of historical buildings and monuments, it includes the re-integration of the urban space into the logic of the urban tissue, and it even crosses the boundaries of the historic city and extends into the periphery to generate places of identity and authenticity. To fulfil this task, we must understand the meaning of the urban form, the structure of the urban space, the relationship between the different components of the urban landscape, and the processes of their transformation. With other words, we must understand the morphology of neighbourhood, district, and city (even region).

Morphological studies have a long tradition in modern science. Introduced in 1790 by the German poet and natural scientist Johann Wolfgang von Goethe as comparative approach to understand the metamorphosis of plants, the method of morphological analysis has been applied and elaborated in many different disciplines, from biology to geography, geology, linguistics, architectural theory, and engineering. Urban morphology has been established as scientific field at the end of the nineteenth century and experienced its first climax with the pioneering work of the Chicago School of Sociology in the 1920s. Since that time, an enormous number of different theories, approaches and methods have been developed, in different countries, from different disciplines, on different scales, which all claim to contribute to the debate on urban morphology.

[1] Manuel Castells, Space of Flows, Space of Places: Materials for a Theory of Urbanism in the Information Age. In: Biskwapriya Sanyal (ed.), Comparative Planning Cultures. New York and London 2005, p. 57.

The broad development of morphological research generated its own limitations. Despite, or just because of the diversity of origins, aims and purposes the debate on urban morphology very soon became canalized in different 'schools', separated scientific 'in-groups' that barely referred to each other and even less communicated with each other. As a result, we are talking about the Italian School, the French School, the Anglo-German School and other schools that behave like scientific 'camps', where the debate on theoretical implications and methodological aspects is being held mainly as internal discussion, not as an open exchange, even confrontation of ideas and approaches between different schools. [2]

Fortunately, this situation has been changed. During the last 20 years, an increasing number of publications appeared that not only aim to explain the differences of approaches and methods and to compare their performances, but also try to combine them in more complex tools of urban analysis. However, most of these publications until now are investigating the possibilities of combined approaches of urban morphology on theoretical level. Empirical studies still are rare, from the methodological point of view often poorly elaborated.

This exactly is the gap that attempts to fill this study. The authors Li Xiaoxi and Zhang Ye place their explorations of urban morphology in the environment of urban planning, design and management by asking what kind of morphological analysis is needed to support the conservation and the management of historical urban landscapes. Based on extensive literature studies, they identify two dominant approaches in urban morphology that together 'enable a better understanding of the historical formation and transformation of urban form': the historico-geographical approach that describes the urban form and its historical origins and changes and the configurational approach that focuses on the relationship between urban form and the use of urban space, the spatial potentiality. Both approaches have been widely elaborated in different analytical tools, but they never have been combined.

Li and Zhang conclude that both approaches are needed to face a specific challenge of conserving and managing historical urban landscapes in recent days: to bridge the tension between the demands of conservation and maintenance on the one hand, the pressure of change and the dynamic of socio-economic developments on the other. An analytical tool is needed that makes it possible to explore the synergies (and contradictions) between cognitive (descriptive) and normative (prescriptive) demands, between the needs of restoration and transformation. Accordingly, Li and Zhang develop a conceptual and a methodological framework to combine both approaches and to generate an integrated approach with the aim to evaluate urban landscape units 'based on the historical expressiveness and spatial potentials of the urban form', capable to 'provide guidance to future management of historical urban landscape regarding to what to change and what to maintain and contribute to a synergy between conservation and redevelopment'.

The integrated approach then is applied in different case studies to test the applicability of the combined method. A first test with focus on historicity is done in

[2] According to Whitehand, the diversity of scientific disciplines and even language problems may have contributed to this situation (Whitehand, 2015).

two cases: in Ludlow, a well-preserved historic town in the UK that is characterized by incremental growth, and in Chinatown, a historic district in Singapore that is used as an example for a fast-changing urban landscape. A more extensive third case study presents a detailed morphological analysis of the Chinese city of Nanjing, with focus on historicity and socio-economic vibrancy in the historic quarter of Nanjing, the Old Chengnan district. As a result, the case studies show that the integrated approach allows a better understanding of the urban landscape, which urban elements are predestinated for conservation and restoration, which elements need physical improvement and redevelopment, how the different landscape units spatially and socially are connected, and how changes to one may affect another. In this framework, Li and Zhang identify important deficiencies in the present practice of urban conservation planning in Nanjing that can be addressed by the proposed integrated approach of morphological analysis.

In summary, the study of Li and Zhang opens a new perspective for the application of morphological analysis in urban planning, design, and management. The integrated approach is innovative and allows a more differentiated understanding of the complexity of the urban landscape than the traditional, more singular approaches of urban morphology. From the methodological point of view, it is carefully elaborated and embedded in the recent debate both on morphological analysis and on conservation planning in the urban context, addressing the contradictory needs of continuity and flexibility in conservation planning in a fast-changing urban environment. On the other hand, it is a first step in the development of more complex analytical tools for urban planning and design. I hope that more studies will follow that widen the application of urban morphology in this field.

Rotterdam, The Netherlands Jürgen Rosemann

Acknowledgements

The seed of this study was planted early in 2013 based on a few inspiring conversations with Prof. Dongqing Han at the Department of Architecture, Southeast University, who unreservedly shared his in-depth knowledge and critical reflection on urban redevelopment in China. It was then gradually developed and enriched by invaluable insights from colleagues and friends in the past years.

First and foremost, our special gratitude and respect go to Prof. Jeremy Whitehand who gave us great encouragement and affirmation when we presented our preliminary work in the 21st International Seminar on Urban Form (ISUF) conference at Porto in 2014. His words had strengthened our faith in embarking on this journey. We would also like to extend our sincere thanks to Dr Fei Chen from the School of Architecture, University of Liverpool, Dr. Kai Gu from the School of Architecture and Planning, University of Auckland, and Prof. Jurgen Rosemann from the Department of Architecture, Delft University of Technology, who have inspired and helped us at different stages of the journey with their great rich experiences and insightful suggestions. Our thanks also go to Prof. Wowo Ding from the School of Architecture and Planning, Nanjing University, Prof. Peter Larkham from the School of Engineering and Built Environment, Birmingham University, and Dr. Vitor Oliveira from the Faculty of Engineering, University of Porto for their constructive feedback on our research.

We would also like to express gratitude to our colleagues at the Department of Architecture, National University of Singapore for their comments at various seminars, from which we benefited immensely in developing our thinking about this study. They are Dr. Yu Chen, Prof. Kai Cao, Dr. Mark Mcdonnel, and Dr. Johannes Widodo.

Also, we are grateful to receive tremendous help from a number of colleagues and friends for data collection and field survey in Nanjing. They are Prof. Li Bao, Dr. Yinan Dong, Dr. Rong Fang, Dr. Xin Ge, Dr. Xinyu Hu, Dr. Lei Jiang, Dr. Xinjian Li, Dr. Haiqian Liu, Dr. Yacheng Song, Prof. Marco Trisciuoglio, Dr. Lian Tang, Dr. Zheng Wang, and Dr. Bing Xia.

Our special gratitude also goes to the two reviewers for their insightful comments on our book proposal and manuscript. We also thank Lisa Fan, the senior editor, Karthikeyan Durairaj and Kavitha Palanisamy, the project coordinators, and Gowtham Chakravarthy V, the project manager from Springer Nature. They have been always patient and kind to grant our numerous requests in the past two years.

Finally, we reserve our sincere thanks to our family members. From Xiaoxi: I owe my husband Tianwei the deepest thanks, who undertook most of the housework during my pregnancy and share most of the parenting load since our son Jiamu was born to let me concentrate on the writing. I would also like to thank my parents and parents-in-law for their enduring love and care for us. From Ye: warmest thanks to my wife and son for their understandings and supports, which were always the source of motivation during the making of this book over the Covid pandemic.

(All names are listed in alphabetical order of the last name in each paragraph).

Xiaoxi Li
Ye Zhang

Contents

About the Authors

Dr. Xiaoxi Li is Assistant Professor in the School of Architecture, Harbin Institute of Technology (Shenzhen), China. She received her B.Arch. and M.Arch. from Southeast University, Nanjing, China. She completed her Ph.D. from National University of Singapore in 2019. She has participated in several urban design and conservation projects in China, during which she developed an interest in exploring evidence-based approaches to urban regeneration. Her current research interests include combining different morphological traditions, and applying morphological methods to urban design and landscape management. She has published a number of papers in the journal of Urban Morphology and international conferences. Her Ph.D. thesis was nominated for NUS Wang Gunwu (WGW) Thesis Award in 2019. She holds membership in International Seminar on Urban Form (ISUF) and serves as Young Ambassador of International Forum on Urbanism (IFoU) since 2020.

Dr. Ye Zhang is Associate Professor at the Department of Architecture, National University of Singapore. He received his Ph.D. from the University of Cambridge (2014) and M.Arch. from Tsinghua University (2008). He was also a visiting academic at The Bartlett School of Planning, University College London (2015). Dr Zhang's principal research interest resides in synergizing different methods for understanding and analysing the built form. He has published widely in the fields of urban form and urban design, and his work has been exhibited in Venice biennale of Architecture, UIA World Congresses, Beijing International Design Week, and many other international exhibitions. Now he serves as the director of Master of Arts in Urban Design programme at the National University of Singapore, and also the Director General of International Forum on Urbanism (IFoU).

List of Figures

List of Tables

Chapter 1
Introduction

Abstract The aim of the book is to understand how the specific historical significance is embedded and contemporary socio-economic activities are organized in urban form and to provide alternative solutions for the conservation and management of historical urban landscape. Three case studies in strikingly diverse cultural contexts are chosen, namely Ludlow - a medieval town in the UK, Chinatown—a historic quarter in Singapore and Old Chengnan—a historic quarter of Nanjing, China. It makes a significant contribution to the growing body of research that coordinates and synthesizes different urban morphological traditions, and to urban planning and design practice as a useful tool for managing changes in historical urban landscape. This book consists of three parts. Part I reviews the relationship between urban morphology and historical urban landscape management. Part II develops the theoretical underpinnings and methodological framework of combining the historico-geographical and configurational approaches. Part III is concerned with applying this evaluation framework in urban conservation practice.

Keywords Historical urban landscape · Urban morphology · The historico-geographical approach · The configurational approach · Urban conservation and management

1.1 Why This Book?

Over the past few decades, Chinese cities have experienced unprecedented urbanization and modernization. The historic inner cities have become the focus of urban development and regeneration processes, experiencing great pressure for change. I have been witnessing to this rapid transformation during my undergraduate study in Nanjing since 2005. One historic area in the old city of Nanjing was chosen as the site of our urban design and regeneration project. Unfortunately, this historically stratified urban landscape, reflecting distinctive residues of past periods, was demolished ruthlessly not long after the design project was completed. The container for the collective memory and identity no longer exists, leaving only street alley names that attempt to remind us of its past history. It is also uncommon to read from newspapers

X. Li and Y. Zhang, *Conserving and Managing Historical Urban Landscape*,
The Urban Book Series, https://doi.org/10.1007/978-981-19-4222-8_1

from time to time about the demolition of another historic urban area to make way for new development in Nanjing.

Not surprisingly, these areas are densely populated residential clusters, characterized with dilapidated traditional buildings and structures, lack of repair and renovation. With the aim to upgrade the living environment with diverse functions and modern buildings, the common practice is restoring the historical monuments and buildings scattered on site while replacing the rest with high-standard housing, international chain shops, hotels, cafes and bars in fake-traditional style catering for tourists and city elites, and relocating the local residents to the outskirts of the city. The historic urban areas have no longer been a place for local people to live in and interact with one another. Instead, they have been gradually capitalized driven by the market economy.

This gave this book its starting point to argue that historic districts could serve as a powerful catalyst for economic and social regeneration, where inhabitants should be able to conduct their lives. Otherwise, there lack essential ingredients of spirit of place. A well-developed and managed historic urban area not only attracts visitors and capital but also local residents as well. The question is how to achieve that socio-economic balance in the regeneration of historic urban areas. This book tries to answer that question by focusing on urban morphology and its application to the conservation and management of historical urban landscape. The notion of historical urban landscape goes beyond traditional terms of groups of buildings, historic centres, ensembles or surroundings, to include broader urban landscape context in this book, which is exactly the research object of urban morphologists. Urban morphology is the study of human settlements, their structure and process of formation and transformation. Our aim is to examine what is really going on in the built environment, especially how the specific historical significance is embedded and contemporary socio-economic activities are organized in urban form. Once we understand how a place works, it is better to apply that knowledge in planning and design to get better results. Thus, the other aim is to put this knowledge into practice, that is to inform more responsive and context-specific conservation and management of historical urban landscapes. It may not be limited to historic cities in China, as the rapid disappearance of historical urban landscapes is also common in other countries. Conservation suggests a careful preservation of particular, while management is defined as a more active process embracing various changes, being less suggestive of restriction to physical preservation in this book.

It starts from combining the historico-geographical and configurational approaches as the first step of combining all the four approaches, as both approaches share the same focus of investigation—the structure of space embodies and explains its socio-economic and cultural processes, and the common object of study—the street—is an essential built form element for both morphological analyses. A theoretical and methodological framework is built up to explore how the two approaches can be used complementarily in both research and practice.

The robustness of this new integrated approach in terms of analysing urban form and the feasibility of its implementation in urban conservation practice are then illustrated by three case studies in strikingly diverse cultural contexts, namely Ludlow -

a medieval town in the UK, Chinatown—a historic quarter in Singapore and Old Chengnan—a historic quarter of Nanjing, China. These cities are representative within their cultural regions. It is found that this integrated framework is capable of providing enhanced understandings of the historical transformations of urban form and cross-scale evaluations of the strengths and weaknesses of landscape units in terms of historicity and spatial potential. This evaluation helps to identify which parts of landscape units exhibit both distinguished historic character and good spatial potential that have comparative advantage for conservation, and which parts confront deteriorated historical expressiveness and poor spatial potential that may need physical improvement for future redevelopment. This could serve as sound reference for planners and designers to manage change, especially which new forms to be incorporated and what functional and locational choices of new developments are suggested. Therefore, this integrated approach could potentially contribute to a synergetic relationship between conserving the historic character and promoting socio-economic development, and eventually sustainable urban planning and design.

This book makes a significant contribution to the growing body of research that coordinates and synthesizes different urban morphology traditions, which has posed a major challenge for this field over the past decade. It attempts to build up the weak relationship between the poorly interconnected strands of morphological research. Besides, by using multi-source data ranging from conventional cartographic maps to computer-generated and emerging open online data, it innovatively relates qualitative and quantitative aspects of urban form, and links the qualitative and quantitative analysis of formal structure. In addition, it makes efforts to remedy the ill-developed relations between researches on different cultural regions.

As an interdisciplinary study merging geography, urban history, urban planning and design, this book is intended a reference guide by graduate students and scholars in various fields who are interested in urban form and urban conservation and management. In addition, it offers practitioners in urban planning and design a useful tool for managing changes in historical urban landscapes. Lastly, it contributes to developing a common platform to facilitate dialogue among various stakeholders and participants in urban conservation practice.

1.2 The Structure of the Book

This book consists of three parts. Part I reviews the relationship between urban morphology and historical urban landscape management. It consists of two chapters. Chapter 2 introduces the different traditions of urban morphology, and its applications and potentials in conserving and managing the historical urban landscape. Chapter 3 brings up the current challenge and opportunities in urban morphology, that is combining and coordinating different morphological traditions. The focus is on the current efforts of combining the historico-geographical and configurational approaches in the studies of the historical transformation and socio-economic performance of urban form. Yet the existing investigations of the complementarity between the two approaches remain not fully clear, which opens up great opportunities for further exploration.

Part II develops the theoretical underpinnings and methodological framework of combining the historico-geographical and configurational approaches with the aim of achieving an enhanced understanding, analysis and prescription of urban form. It consists of three chapters. Chapter 4 builds up the theoretical and analytical frameworks that combine the two approaches. Chapter 5 explains how this integrated framework could help understand the historical development of urban form by using Ludlow, a medieval town in the UK as an example. This positive result of the first case allows us to apply this framework in a drastically different urban context to further test and demonstrate its robustness, that is Chinatown, a historical quarter in Singapore. Moving on from the historical expressiveness of urban form, Chap. 6 examines its contemporary socio-economic performance under this integrated approach. These two aspects are then incorporated into a single framework to evaluate the strengths and weaknesses of landscape units across different scales, which provide implications for future improvement. A detailed case study of Old Chengnan, a historical quarter in Nanjing, is carried out to test this framework.

Part III is concerned with applying this evaluation framework in urban conservation practice and landscape management. It is comprised of three chapters. Chapter 7 introduces the current conservation planning policy and conservation practice in China. Nanjing as one of the first-designated historic and cultural cities is chosen as an example. Chapter 8 elaborates how this newly developed framework could contribute to area-based urban conservation in terms of delimiting boundary and establishing management priority by comparing the conservation proposal prepared by the integrated approach and that by the planning authority. Chapter 9 reflects on the key findings and significances of this research, as well as the limitations and recommendations for future research.

Part I
Context and Theory

Chapter 2
Urban Morphology and Historical Urban Landscape Conservation and Management

Abstract This chapter introduces the main morphological approaches that have been developed since the 1960s: the historico-geographical approach after the seminal work of M. R. G. Conzen, the process-typological approach following the morphologist—Saverio Muratori, the configurational approach developed by Hillier and Hanson and the spatial analytical approach. The theoretical and methodological framework of each morphological approach is introduced as well as its applications in the urban planning and design practice. Meanwhile, there are widely spread problems in the historical urban landscape management, that is the lack of a systematic method of delimiting the areas to be conserved and building up management priorities. Urban morphology has the potential of an interdisciplinary field of study to contribute to the understanding and management of historical urban landscape in a period of unprecedented change.

Keywords Urban morphology · Fast-changing urban landscape · Conservation and management

2.1 Urban Morphology

Urban morphology as an interdisciplinary field of knowledge (geography, urban and architecture history, archaeology and urban planning) has grown rapidly in recent years. It is the study of human settlements, their structure and their process of formation and transformation. The aim of the morphological study is to contribute to the understanding of the built environment as a complex physical object (Kropf, 2018). There are four approaches to urban morphology: the British school initiated by Conzen, the Italian school established by Muratori and Caniggia, the French school started by Panerai and Castex in the 1960s, and the space syntax approach explored by Hillier in the 1980s.

Kropf (2009) has undertaken a critical analysis of different approaches to urban morphology. In his framework, the British school is referred to as the historico-geographical approach, the Italian school as the process-typological approach, the space syntax approach as one representative of the configurational approach, for an easier recognition and comparison with other morphological approaches. The French

X. Li and Y. Zhang, *Conserving and Managing Historical Urban Landscape*,
The Urban Book Series, https://doi.org/10.1007/978-981-19-4222-8_2

school is not highlighted, as the work is not accessible to an English-speaking audience compared to the other three widely acknowledged schools (Moudon, 1997). The spatial analytical approach is included for its ability to provide a description of the blurring aspects of form and use at different spatial scales through a range of models (Kropf, 2005, 2009, 2018). Each morphological approach captures and elaborates different aspects of urban form: the process-typological and configurational approaches have origins in the fields of architecture and urbanism, while the historico-geographical approach and spatial analytical approaches are from the field of geography (Kropf, 2017). In the following sections, the concepts, methods and applications of these four morphological approaches are elaborated separately.

2.1.1 *The Historico-Geographical Approach*

The historico-geographical approach originated from the German historical geographers at the end of nineteenth century and developed in the 1960s. It is also known as the 'Conzenian school', after its principal originator, M. R. G. Conzen. This approach holds the view that a town is subject to change and its development is written deeply into the outline and fabric of built-up areas. Thus tracing existing forms back to the formative processes is an important means of learning from the success and failure of past societies in their attempts to shape the built environment (Conzen, 1960).

To disaggregate urban landscape into its component parts, a tripartite division of form - ground plan comprising street, plot and block plans of building, land and building utilization, building fabric was put forward by Conzen (1988) as 'form complexes' (Fig. 2.1). Among the three form complexes, their degrees of resistance to change contribute to the hierarchy of landscape units. Ground plan (plan unit) as the most resistant to change determines the top rank; land and building utilization tends to be more ephemeral, thus lies in the intermediate rank; buildings are on average, that partakes in the intermediate rank but makes its major contribution in the lowest rank (Table 2.1).

Integration of the maps of the three form complexes yields a multi-tier hierarchy of morphological regions (landscape units), that is an area of homogenous urban form in terms of plan type, building type and land use (Conzen, 2004). Morphological periods reflect 'socio-economic and political changes linked in varying degrees

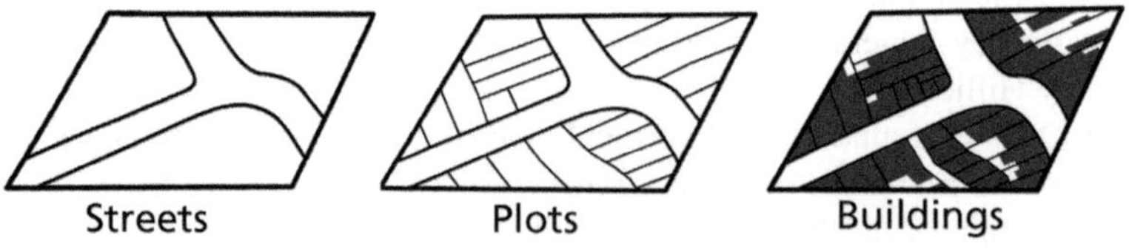

Fig. 2.1 Three form elements of ground plan (plan unit) (Kropf, 2018)

Table 2.1 Contribution of different form complexes to urban landscape characterization (Conzen, 1988)

Attribute	Persistence	Contribution to hierarchy
Ground plan	High	Mainly high and intermediate rank
Building fabric	Variable but often considerable	Mainly intermediate and low rank
Land and building utilization	Low	Mainly low and intermediate rank

to the introduction and spread of a vast number of artistic, technical and other innovations' (Whitehand, 1977). They are a major feature of additions to new forms and internal changes and replacements of existing forms in cities. The change of three form complexes combined with distinctive morphological periods is evident to uncover how urban landscape is historically stratified. The first-order boundaries reflect the main stages in the historical development. The second- and third-order boundaries reflect plan changes within first-order plan units caused by subsequent redevelopments, which could be boundaries of land use or building form. Landscape units of the lowest order capture the variations of building forms or minor differentiations of plan (Conzen, 1988).

Ground plan is the most resistant to change among the three *form complexes* (ground plan, building fabric and land use); it encapsulates the most information of the historical development and transformation of a town (Conzen, 1960). Ground plan analysis provides an important basis for delimiting morphological regions (landscape units). The seminal work in ground plan analysis is Conzen's study of the evolving town plan of the English market town of Alnwick (Conzen, 1960, 2004). The analysis was at various resolutions from the entire town to individual plots and buildings. The hierarchical mapping shows how the layout of the town had come into existence and changed over time from early-modern, to industrial and twentieth-century periods (Fig. 2.2). The first-order divisions (denoted by capital letters) distinguished the old town and new accretions. The second-order divisions (marked by Roman numerals) identified the areas with relative internal homogeneity, such as the old town, the residential accretions and fringe belts which disrupted the built-up areas. The third order (denoted by small letters) distinguished the subdivisions within the second-order units, such as the kernel of the old town, the intramural and extramural areas of the fringe belt. The fourth order (marked by Arabic numerals) was distinguished by plan type, such as the ancient deep-burgage series, traditional arterial ribbons and so on. Following Conzen's detailed plan analysis of Alnwick (1960), the historico-geographical approach has been widely applied to understand the origin and development of British historic towns, such as Newcastle upon Tyne (1964) and Ludlow (1988), through mapping the morphological evolution of the town plan in different periods (Baker & Slater, 1992; Lilley, 1998, 2000; Slater, 1990). In the past 20 years, there have been applications of Conzen's method of plan unit in other types

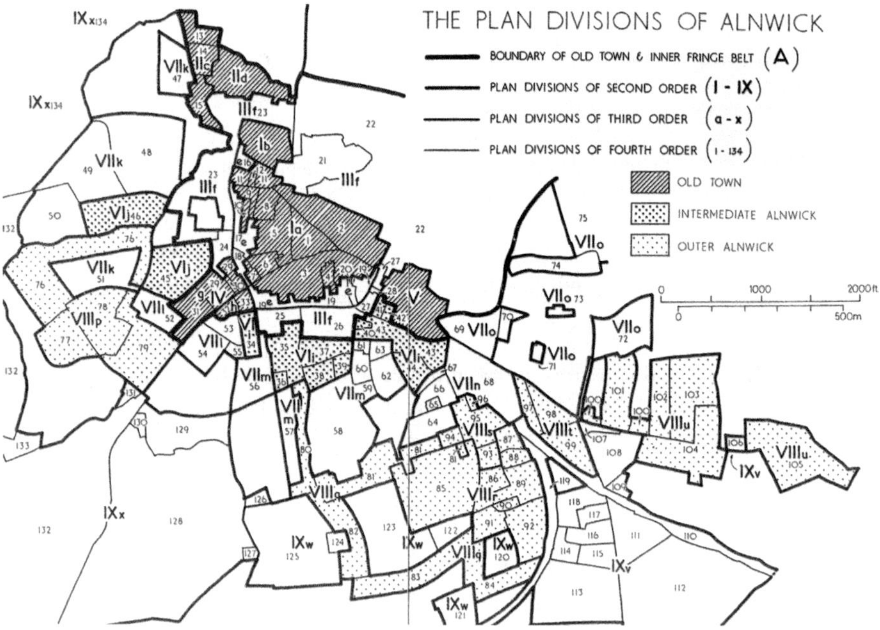

Fig. 2.2 A four-tier plan unit of Alnwick, reflecting the historical development of the town (Conzen, 1960)

of urban areas and other parts of the world, including Asian cities (see, Gu, 2001; Whitehand & Gu, 2007a, 2007b, for example).

Moving on from the ground plan analysis, Conzen later delimited the morphological region (landscape unit) in a detailed study of the English market town of Ludlow (1988). Ground plan, building type and land use utilization were divided into different hierarchies and integrated to produce the landscape unit division (Fig. 2.3). A five-tier hierarchy of landscape units was recognized that reflected the historical development of the town, ranging from the entire town at the top of the hierarchy to the 'smallest coherent form associations' at the bottom hierarchy. Conzen's conceptual framework and methodological thinking are remarkable in the field of historical urban morphology. It exhibited the sharpness of plan unit or landscape unit boundaries that reflect the historical development of urban areas. Due to the different values and opinions held by researchers, boundary division of plan unit or landscape unit may not be replicable. This debate was acknowledged by Whitehand (2009) that "it would be unrealistic to expect… patterns of urban landscape units to be precisely replicated by researchers". Yet the subjective character of boundary delimitation could be reduced by a clear concept of reasons in each case to show why they are appropriate and defensible, if not replicable by different researchers.

Fig. 2.3 Landscape unit division of Ludlow in 1946 (Conzen, 1988)

Because of its focus on the process of formation and transformation of urban form, the historico-geographical approach is able to recognize, delimit and manage historical urban landscape (Kropf, 2017; Whitehand, 2015). By articulating the historico-geographical structure of urban landscape, it provides on the one hand a more rational basis for delimiting areas for conservation, on the other hand, sound evidence for proposing areas to change and incorporating new forms in old landscapes. There have been a growing body of research on the applications, adaptations and demonstrations of this approach in the UK and in China, especially in delimiting the boundaries of historic areas and identifying conservation priorities (Baker, 2010; Kropf, 2011; Larkham, 1996; Thomas, 2018; Whitehand & Gu, 2007b; Whitehand et al., 2011). Opportunities for area enhancement (e.g. Baker, 2010; Whitehand et al., 2011) are proposed mainly based on historical character assessment, which provides guideline for future development to maintain the character and identity of an area in both landscape management and conservation planning.

Historic England on Conservation Area Appraisals (2011) has undertaken *historic urban characterization* as a way of identifying and understanding the characteristics of historic urban areas (Kropf, 2011; Thomas, 2006). Rather than just an area which is 'desirable to preserve or enhance', characterization defines the 'particular characteristics of each area and sub-area-its component parts, structure, function and origins' (Kropf, 2011, p. 395). A fine example of this historico-geographical

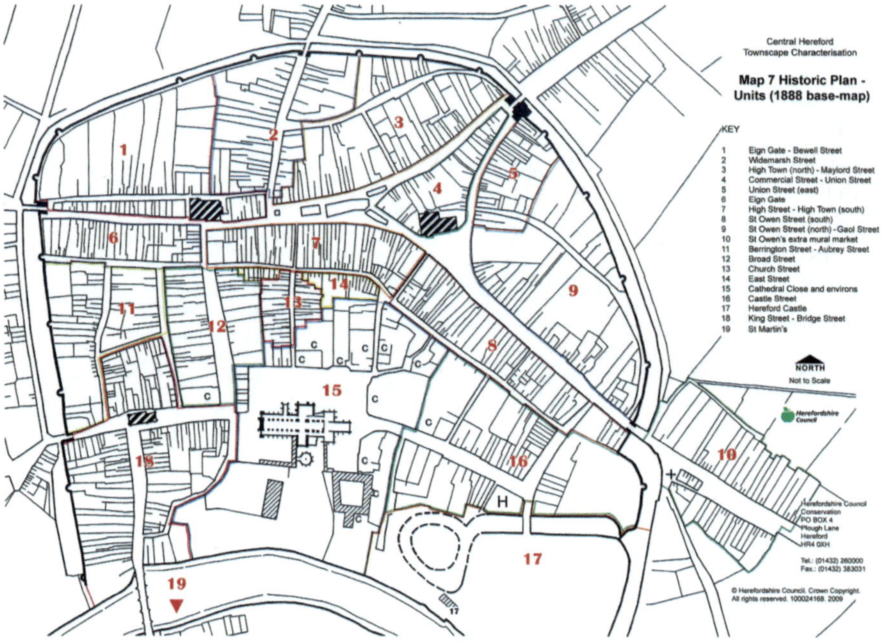

Fig. 2.4 Plan unit division of Hereford town in 1888 (Baker, 2010)

approach is the study of Central Hereford (Baker, 2010) commissioned by English Heritage with the aim to examine what its historical components are, their state of preservation and their significance. Plan unit divisions (Fig. 2.4) reveal the origin and development of the town as well as its component parts. By integrating land use with historic buildings and plan unit, the character areas of the town were revealed (Fig. 2.5). The output consisted of a map of character areas with written descriptions of the value and significance of each subarea as well as of the opportunities for future regeneration, enhancement and management. This study was designed to inform the Conservation Area Appraisal by Herefordshire Council in the longer term.

The historico-geographical approach has also demonstrated its power in providing an understanding of urban areas as historico-geographical entities and strengthening the theoretical and methodological basis of historical urban landscape management, through a systematic examination of the morphological characteristics of Chinese cities. Take Beijing's Zhishanmen historic urban area for example, the map of urban landscape units delimitated by the historico-geographical approach provide the basis for conservation proposal. It differs from the conservation area drawn by the planning authority which individual sites and buildings are treated in isolation with the wider historico-geographical environments (Fig. 2.6). Moreover, the historico-geographical analysis provides guidance for future development in each morphological region (landscape unit). For instance, within the high-density courtyard unit, the buildings and plots facing the main streets need to be sensitively adapted and updated in accord with the character of this landscape unit and avoid overcommercialization or tourism. At the same time, the introduction of new designs shall be sympathetic to

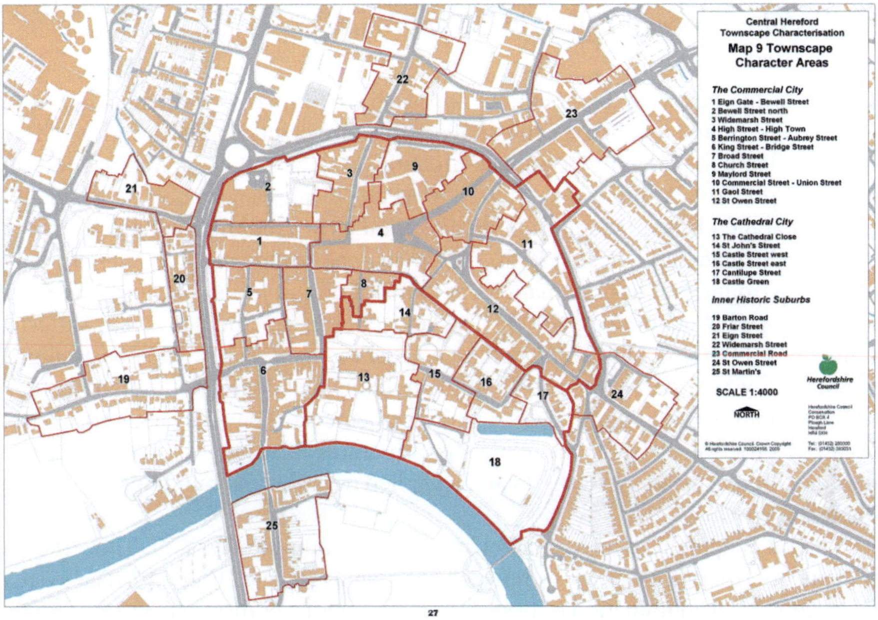

Fig. 2.5 Landscape unit division of Herford town in 2010 (Baker, 2010)

the old in building style, plan and height. This case study successfully demonstrates that an understanding of urban areas as historico-geographical entities can strengthen the theoretical basis of conservation in very different cultural conditions.

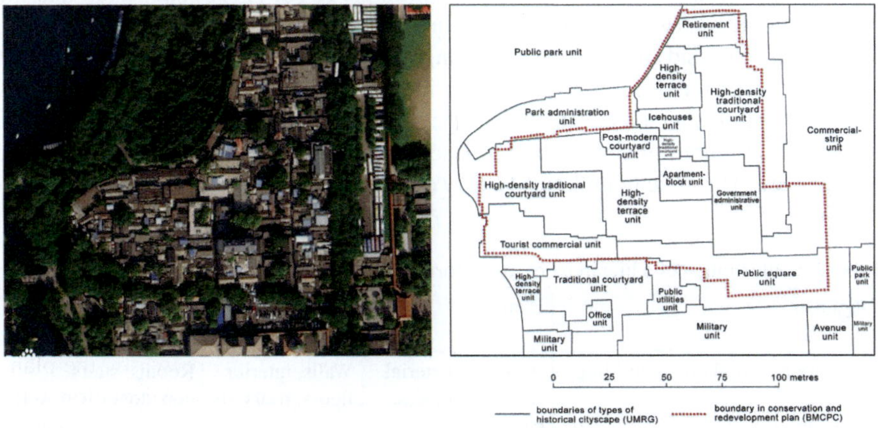

Fig. 2.6 Left: the Google map of the Zhishanmen historic urban area. Right: comparison of boundaries drawn by Beijing Municipal City Planning Commission (red dotted line) and by the Conzenian approach (black continuous line) (Whitehand et al., 2007b)

2.1.2 The Process-Typological Approach

Parallel to the historico-geographical approach, the process-typological approach began to take shape within the context of architecture practice and education in Italy in the 1960s. It was mainly developed around works of architects Muratori (1959) and Caniggia (1982) and is also referred to as the 'Caniggian school'. Caniggia has transmitted the ideas of Muratori in architectural terms, highlighting its more operative aspects than theoretical systems. A typological method capable of interpreting human environmental transformations in terms of design for architectural purposes was developed (Cataldi et al., 2002). The key concepts and methods are manifested in the book *Composizione architecttonica e tipologia edilizia* written by Caniggia and Maffei, which starts at the scale of the building and develops through aggregations of buildings as urban tissues to entire settlements. The analysis is based on a hierarchy order of subdivisions: elements, structures of elements, system of structures and organisms of systems. It can be applied to individual buildings or to towns (Table 2.2).

When it is applied to towns, the elements are buildings (Fig. 2.7). The structure of elements is urban tissue, also referred to as an aggregate (Fig. 2.8). The system of structures is a combination of tissues forming regions or districts, which form the organism of the town. The forms found at different levels are identified as types which are conceived as cultural entities rooted in the local process of cultural development. The aim is to turn the current architectural crisis into operational awareness so as to understand its historical roots (Caniggia & Maffei, 2001).

This approach seeks to inform architectural and urban proposals with an understanding of the built environment by examining its detailed structure and historical process of its formation. The notion of urban tissue helps to understand the relation between urban areas and individual buildings. Kropf (1996) suggested that the process-typological approach is effective in informing form-based zoning in planning at building, plot and district scales. A range of types that is a product of a distinct social, historical and cultural process is identified. This identification provides a firm basis for character assessment and the definition or review of conservation area boundaries. Figure 2.9 shows the proposed form-based zoning plan for the centre of Mennecy. The zone boundaries largely correspond to the tissues identified by the

Table 2.2 Illustration of a hierarchy of urban form analysis at the scale of building and town

Organism		Element	Structure of element	System of structures
Resolution	Individual building	Building material (Bricks, timbers, etc.)	Walls, interior floors, roofs	Rooms, stairs, corridors
	Town	Building	Urban tissue	Regions or districts (a combination of tissues)

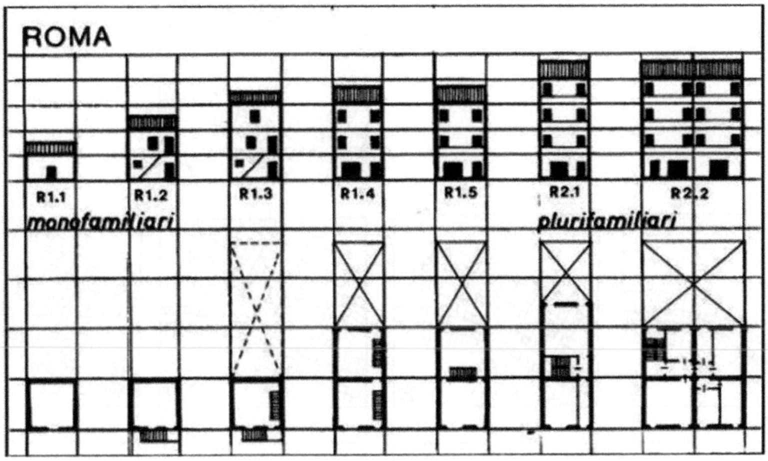

Fig. 2.7 Typological transformation of dwellings in Rome (Caniggia & Maffei, 2001)

typological analysis. The zones also indicate the building guidance for future development in the area. For plot in each zone, the regulations are generated including the frontage dimension and the maximum area, buildable area, the arrangements of buildings and the type of buildings (corresponding to component parts). This typological study is an alternative approach to development control within the French local land use plan and provides great potential to fit into the planning system with the purpose of maintaining local and historical character of the individual areas and the town as a whole.

The concept of 'typological process' didn't appear until the 1970s in the particular Italian context. The English translation of the substantial part of the book *Composizione architecttonica e tipologia edilizia* (*Architectural composition and building typology 1: reading of basic building*) in 2001 was a resurgence in urban morphological research and especially its communication between researchers. It still poses challenges today, both theoretical and empirical, and is stimulating new interdisciplinary and cross-cultural research (Gu et al., 2008; Li & Gauthier, 2014; Whitehand et al., 2014).

2.1.3 The Configurational Approach

2.1.3.1 Space Syntax

Divergent from the historico-geographical and process-typological approaches, the configurational approach, especially space syntax, was following the lines of mathematical and quantitative investigations of architectural and urban form in the 1960s by Martin and March in the UK (1972), and was developed by Hillier and Hanson

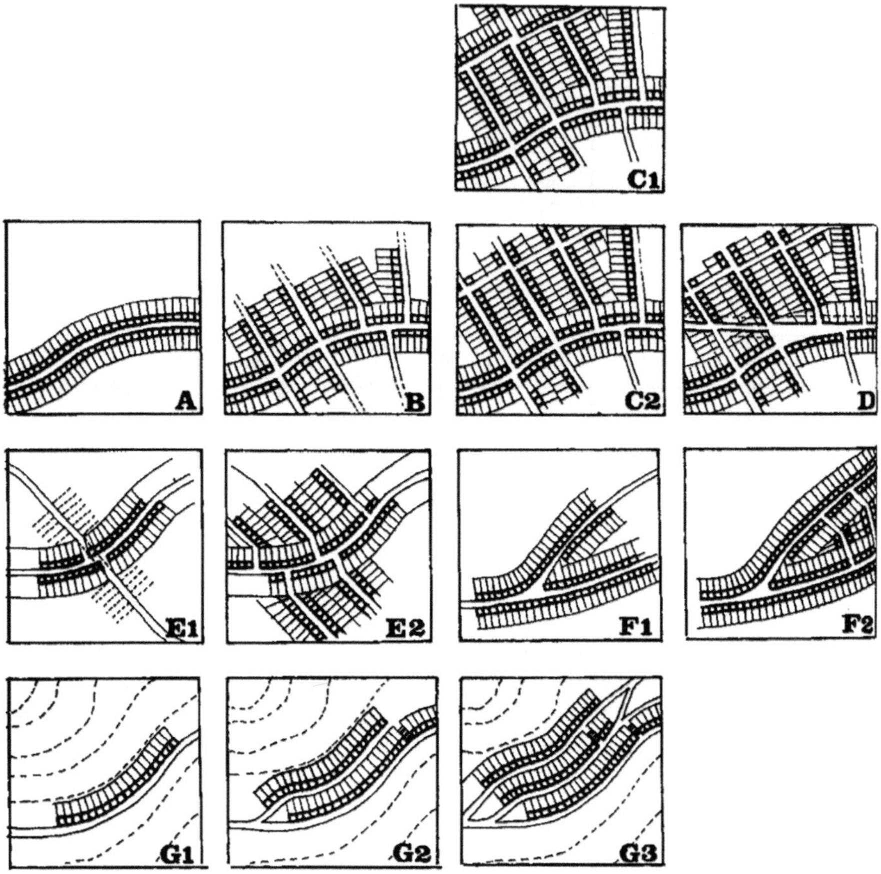

Fig. 2.8 Formation of urban tissue or aggregates in different generalized situations. A: building along matrix route, B: building along planned building route, C1: formation of connection route between planned building routes with the subsequent formation of planned building routes continuation, C2: formation of connecting routes between planned building routes subsequent to the extension of planned building routes, D: break-through street formation. E1, E2: building along matrix route and along derived planned building routes in the event of intersection between two pre-existing pure routes. F1, F2: building along matrix route and along derived planned building routes in the case of forking of pre-existing pure route. G1, G2, G3: building along matrix route and triggering off of planned building routes in steeply sloping area (Caniggia & Maffei, 2001)

(1984). It mainly focuses on the geometric and topological structures of urban form, exploring the relations between urban streets by taking into account other relations in an overall street structure, and is called spatial configuration. A series of representation and modelling techniques have been developed to analyse the configuration of street network, which is also known as 'spatial accessibility'. Accessibility is widely discussed in geography and transportation science, and it is a measurement of 'the spatial distribution of activities about a point, adjusted for the ability and the desire

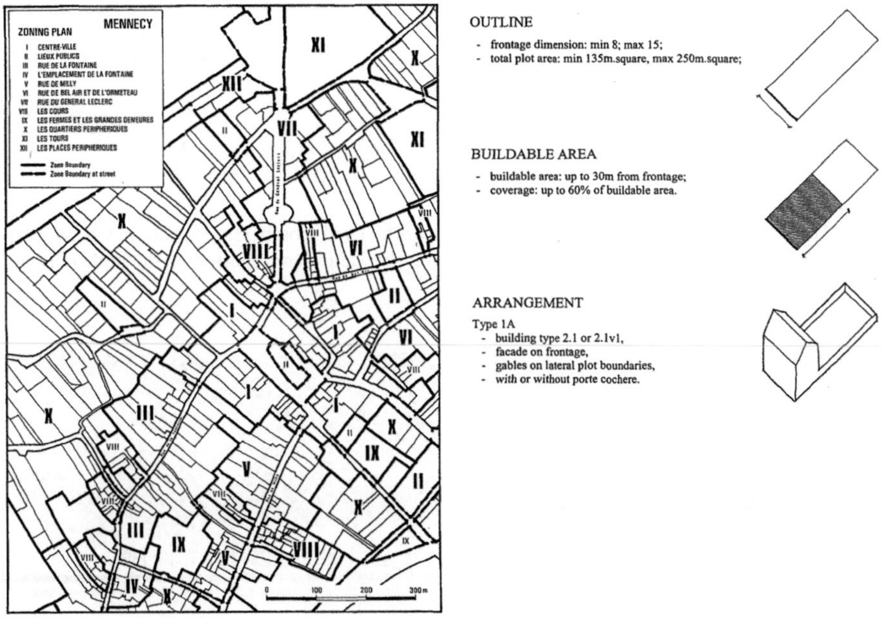

Fig. 2.9 Left: proposed zoning plan for the centre of Mennecy. Right: example of regulations for plots from the Centre Ville zone. Left: examples of regulations for plots from the Centre Ville zone (Kropf, 1996)

of people to overcome spatial separation' (Hansen, 1959, p. 73). Spatial accessibility is a special case of accessibility, which utilizes graph theory to measure the spatial separation of street network. All destinations are accounted as equal and land use is not considered (Batty, 2004; Charalambous & Mavridou, 2012).

Graph theory provides a means of analysing the structure of relationships between elements. There are two types of network elements—nodes and edges. Street networks are represented in two ways: edges as junctions, and nodes as street segments; edges as street segments, and nodes as junctions (Fig. 2.10). The former is known as dual representation while the latter is primal representation. These two representations explore the degree to which an edge or a node is spatially connected to the surrounding network.

There are empirical findings that the correlation between spatial configuration and urban movement underscores a number of economic and social urban phenomena (Fig. 2.11). This gives rise to the theoretical assumptions of 'natural movement' (Hillier et al., 1993, p. 32) and 'movement economy' (Hillier, 1996, p. 54). The configurational approach (space syntax) has been proven effective in identifying lively local centres which are socially and economically viable (Yang & Hillier, 2007), based on the relationship between properties of street network structure, movement and land use.

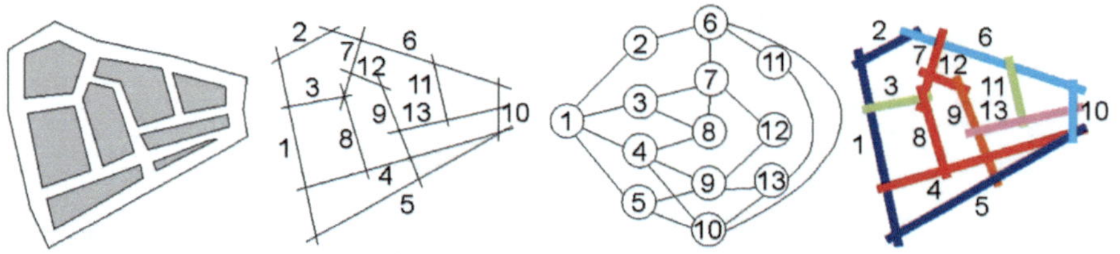

Fig. 2.10 Representation modelling of street network. Dual graph representation where street segment is the node and the relationship is the edge. The values are reported on the node (Crucitti et al., 2006)

Fig. 2.11 Relationship between spatial accessibility and land use. Left: spatial accessibility in the city scale of London. Right: land use distribution of London (Hillier, 1996; Hillier et al., 1993)

To measure how each street line is connected to others that it intersects, angular segment analysis is more frequently used currently. It leads to the possibility that the road centreline presents the most accurate angular change in urban space, and therefore is potentially the most accurate model for movement based on angular change. This segmentation allows finer scales in depicting urban structure than axial lines, by measuring each segment individually (Dhanani et al., 2012; Hillier & Iida, 2005; Hillier & Penn, 2004) shown in Fig. 2.12. It also minimizes the effects of cartographic difference between representations put forward by Ratti (2004).

New advancement on angular analysis was introduced by Hillier, Yang and Turner (Hillier et al., 2012), that is the normalization of angular integration (NAIN) and choice measures (NACH). The purpose was to enable the comparison between systems of different sizes. Normalized segment angular choice (NACH) turned out to be as powerful as integration in both analysis and movement prediction and provides a deeper understanding of the spatial morphology of cities compared to previous least angle integration or choice measures.

The combination NACH and NAIN values provides a framework for explaining the dual structure of foreground and background networks, which is called *star models.* The maximum NACH and NAIN values are able to highlight the foreground structure, while the mean NACH and NAIN values capture the background structure. The maximum value indicates how structured a city is. Mean values are indicative of

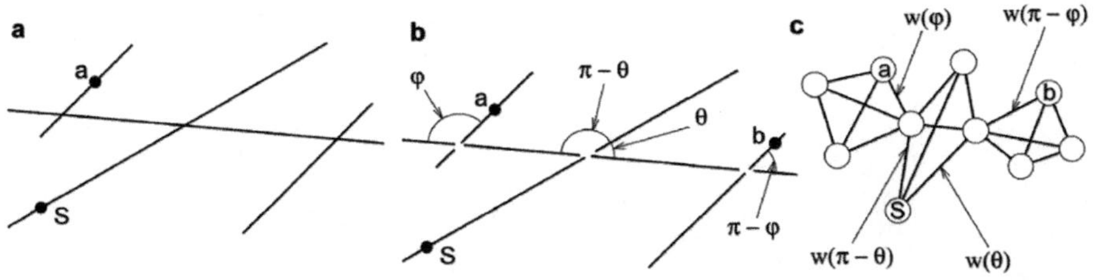

Fig. 2.12 Topological distance in the axial analysis and angular distance in the segment angular analysis. **a** Axial line. The topological depth between a and s is 2; **b** segment line; **c** graph representation of segment lines. The distance between street segments is calculated by weighted angular degrees by Tulip analysis, corresponding to people's spatial navigation and wayfinding in the city. The distance between a and s is calculated as $w(\pi - \theta) + w(\varphi)$, when $w(0) = 0$, $w(\pi/2) = 1$ (Hillier & Iida, 2005)

how grid-like a city is. Maximum and mean values of NAIN are related to the ease of accessibility in the street network. Maximum and mean NACH are indicative of how deformed or interrupted the foreground or background street network is.

The star model could be used to discuss and compare the global forms of individual cities. Figure 2.13 shows two examples out of the fifty cities investigated by Hillier. Based on their differences, tentative interpretations of the underlying forces (either economic or social) in shaping urban forms were made. In Manhattan, it is suggested that the strong background network tends to minimize the difference between it and the foreground network, so as to equalize economic opportunity. Meanwhile, Beijing has a strong hierarchical foreground, so as to create a spatial dominance of the background network. While Manhattan is more top-down economic driven while Beijing top-down socially driven (Hillier et al., 2012).

Another new development was the definition of local structures, with the hypothesis that the local identity of a city is rooted to some extent in the spatial configuration of the local physical environment. The local structures are patchworks of metrically similar areas in the background network, suggesting locally walkable areas (Hillier et al., 2007). These could be identified by simply calculating the metric mean depth from segments within a metric radius. Metric mean depth is one of the two ways of metric weighting, that is to weight the relation for the metric length of the segment (line), so in effect substituting a measure of the length of the segment for the topological value of 1. However, metrically weighted choice measures have been argued to have far less ability in prediction of movement and functional patterns (Hillier et al., 2007).

$$\text{MMD} = \text{Total metric length}/(k - 1), \ k \text{ is the number of segments}$$

The empirical explorations show that patchwork patterns correspond with localities known to have socio-economic and culture characteristics, in either deformed cities such as London, or regular grids such as Manhattan, Barcelona and Atlanta

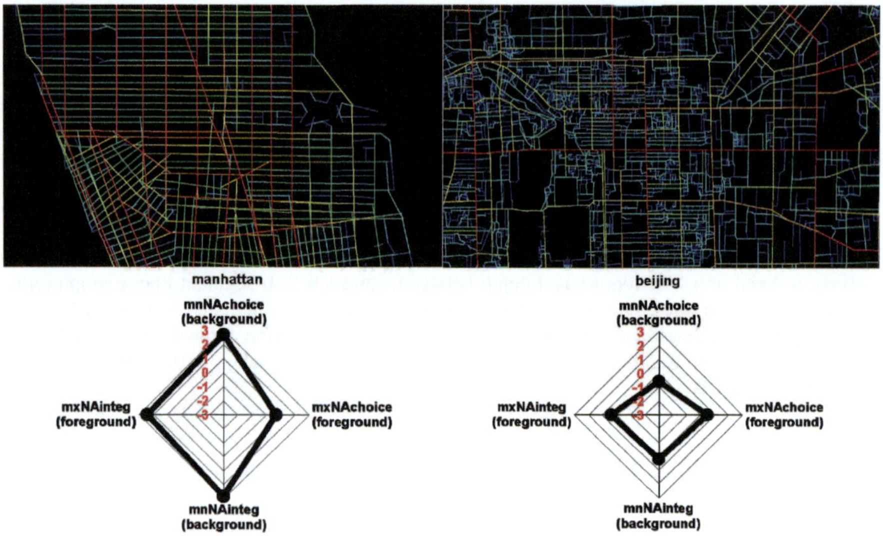

Fig. 2.13 Star models of Manhattan and Beijing. Left: Manhattan has a uniform grid. It is very strong on movement in the background network, but relatively weakly structured in the foreground network. Right: In contrast to Manhattan, the background areas of Beijing are comparatively isolated from each other by the foreground grid, giving a strikingly low background value compared to Manhattan (Hillier et al., 2012)

(Al-Sayed et al., 2014; Yang & Hillier, 2007). Figure Fig. 2.14 displays the patchworks identified at a varied radius of 500 m and 2 km, in which the larger radius identifies larger patches. It shows that there seems to be a correlation between the patchworks and functional variation in the city.

The above discussions show that an analysis of the topological, geometric and metric properties of the street structures helps us to acquire a better understanding of cities. It exposes some of the processes that lead to the formation of urban grid and the role of different social and economic forces in shaping urban form, based on the extensive studies of hundreds of cities. Recent space syntax research has also emphasized that spatial networks of connected space, especially its formative process, could potentially elaborate urban formation and transformation (Griffith et al., 2012; Karimi, 1997, 1999) and inform conservation area delimitation (Palaiologou & Griffiths, 2019).

In Karimi (1997)'s study, Kerman as an Iranian city has undergone great transformation under modernization. The historic urban centre became decayed and unattractive, which led to great concerns about the preservation of historic cities. The configurational analysis reveals that the historic centre of Kerman becomes much more segregated globally because of the construction of modern streets. In Fig. 2.15, the comparisons between 1800 and now show that traditional elements of the historic centre like bazaars, mosques and shrines used to be the most integrated, yet they now rank the lowest. This provides an explanation for the seclusion of traditional elements

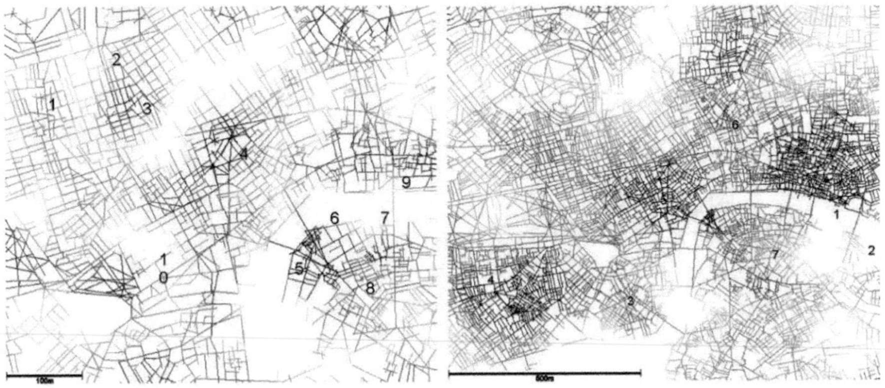

Fig. 2.14 Patchworks of Central London at the radius of 500 m (left) and 2 km (right). Left: 1. Immediate area of Marylebone High street. 2. The adjacent residential and non-live business areas of north Marylebone. 3. Live area of Goodge street. 4. Live Coptic street area south of the British museum and the adjacent live Seven Dials areas. 5. South Bank Cultural centre. 6. Adjacent Coin street. 7. Successful urban regeneration. 8. Roupell street residential area. 9. Live St Andrew's Hill local area. 10. Upmarket non-live area of St James. Right: 1. London's financial centre and its northern business extensions. 2. Area around Borough Market, a highly active regenerated area. 3. Pimlico. 4. Upmarket area. 5. Trafalgar square. 6. Gentrifying area of Clerkenwell. 7. Popular and active area around the south of river (Hillier et al., 2007)

after modernization, as well as the decline and unattractiveness of the historic centre. The findings suggest that the fate of the historic core is strongly dependent on the way its spatial organization is transformed. The inconsiderate transformation of urban grids might undermine spatial organization and consequently lead to a deterioration of urban functions in historic town centres. It is further argued that urban conservation shall be more conserved about the 'spatial spirit' of the place instead of restoring the individual buildings or spaces.

Space syntax's popularity has continuously grown in recent years. An ever-growing number of international scholars and practitioners are applying space syntax at various scales, from neighbourhoods to metropolitan areas and entire regions. It shows how urban phenomena can be understood through configurative studies connected to people's socio-economic behaviours (Hiller & Vaughan, 2007; van Nes & Yamu, 2021).

2.1.3.2 Place Syntax

Space syntax measures the spatial integration–separation of street network, that is also known as 'spatial accessibility'. Street is the study focus, and it is one of the form elements that connect building and plot (Marshall, 2009). Place syntax brings attraction into calculation, which measures the 'spatial distribution of activities about a point, adjusted for the ability and desire of people to overcome spatial separation' (Hansen, 1959). In other words, it measures the accessibility within the street network

Iranian Old Cities

City Elements	*average int. Rn*		
City Square	1.3670	10 Mosques	1.0939
3 Carvansaraye	1.3546	11 Local Bazaars	1.0773
4 Grand Mosque	1.2696	mean integration	1.000
5 Friday mosque	1.2679	12 Shrines	1.0619
6 Colleges	1.2588	13 Castle, Citadel	1.0358
7 Govern. Place	1.1970	minimum integration	0.5543
8 Ruler's Palace	1.1647	mean integration	1.000
9 Local Squares	1.1095	14 Gates or Entrances	0.9723

		9 Hospitals	1.552
1 Cinemas	1.775	10 Grand Mosque	1.415
2 City Hall	1.743	11 Shrines	1.274
3 Banks	1.732	12 Friday Mosque	1.233
3 Guild Hall	1.527	13 Schools	1.223
4 Shopping Streets	1.713	13 Public Offices	1.370
5 Libraries	1.644	14 Bazaar	1.172
6 Museums	1.594	15 Carvansaraye	1.151
7 Fire/Police Station	1.577	16 Local Squares/Bazaar	1.017
8 Public Offices	1.570	17 Mosques	1.007

Fig. 2.15 Overlapped traditional elements with global integration map of the historic core in Kerman in 1800 and today. The average integration values of each element put in a rank order. In 1800, the bazaar, mosque and shrines were mainly located along the most integrated streets in the city, while now these elements are ranked the lowest. (Karimi, 1999)

to different forms of attractions that could be either variables of spatial form like number of plots or human activity like residence or retail (Fig. 2.16). This approach could also link the street network structure with non-formal factors, like the number and differentiation of people and things, such as plot number, amenities, population and so on, analysing how socio-economic activities are organized in the built environment and what are the potentials (Bobkova et al., 2017; Marcus, 2007, 2010; Marcus et al., 2017). This allows us to better understand the interaction between urban form and human activity, that is how urban form supports, impedes and organizes vibrant urban life.

'Attraction reach' as one of the indices in place syntax calculates the number or the sum of the value of a specific attraction within a specific radius via the street network. This indicates the capacity of spatial form to generate and support such attractions. Take the empirical study of Sodermalm in Stockholm for example (Marcus, 2010): Fig. 2.17 (left) shows the conventional representation of building density per plot (geographical unit), while Fig. 2.17 (right) deals with representations of building density within a certain walking distance. It shows that accessible building density is more concentrated around the centre of the city where building density is high and the network dense. It could be concluded that building density could be understood through the lens of accessibility. Further, the correlation between accessible building density and accessible population density was around 82%, indicating that place syntax analysis is powerful in revealing how urban form is influential on urban life.

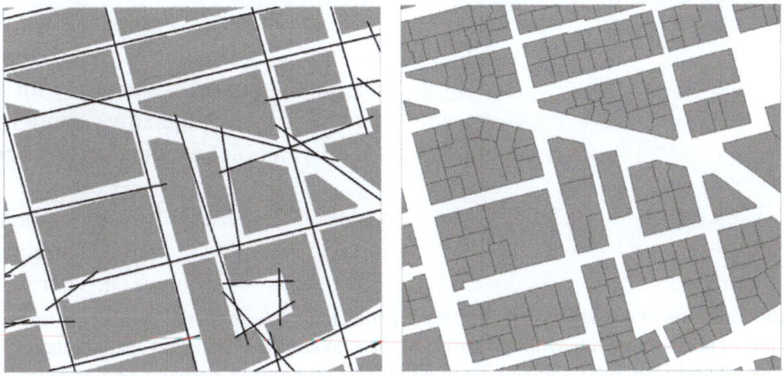

Fig. 2.16 Experientially defined space, where each axial line represents a space that is visually and physically accessible (left), and legally defined space, where each plot represents a domain of an actor, where attractions loaded as place data on either plots or address points (right) (Marcus, 2005)

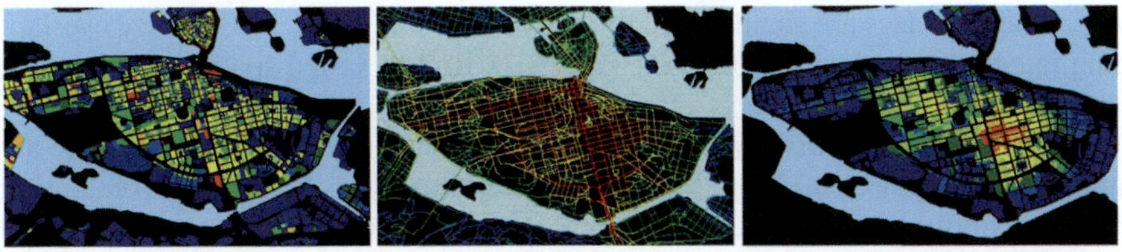

Fig. 2.17 Representation of accessible building density in Sodermalm. Left: building density measured at the unit of plot in Sodermalm. Middle: Global accessibility. Right: accessible building density within a certain walking distance radius, which correlates to accessible population density by 82% ($R^2 = 0.82$) (Marcus, 2010)

Researchers like Jacobs (1961) have long argued against mono-functional districts, and they claim that we live in cities so that we can get closer to many different things. A mixture of primary uses within each district and short blocks and multiple intersections are prerequisites for creating urban diversity. This statement has been widely accepted and used in contemporary urban discourse (Stahle, 2016; Mashhoodi and Pont, 2011). Place syntax could measure such diversity by taking account of a spatial dimension. Empirical studies (Bobkova et al., 2017; Marcus, 2005, 2007, 2010; Marcus et al., 2017; Stahle et al., 2005) have found that spatial accessibility and spatial diversity (accessible plots within a certain distance) are correlated with diversity indices like amount of population and business, which indicate that variables of urban form that can potentially contribute to urban diversity. Based on the findings, the concept of 'spatial capital' is developed to describe the potentials of place to condition socio-economic activities accessible to a specific location (Marcus, 2005). This spatial capital presents the opportunity to discern urban socio-spatial categories, such as super-urban (high accessibility, high diversity); pseudo-urban (high accessibility, low diversity); suburban (low accessibility

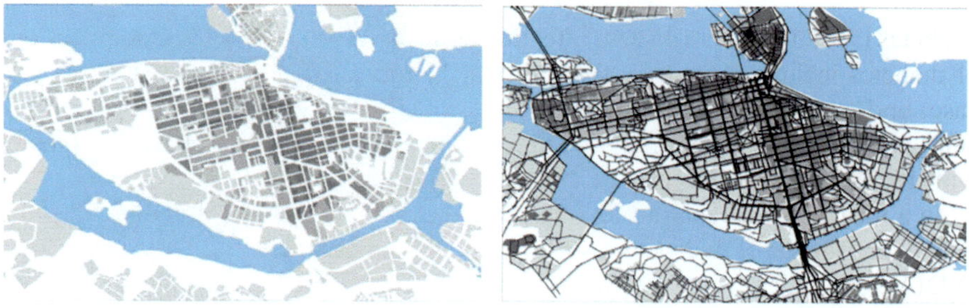

Fig. 2.18 Variation of spatial capital as accessible diversity, where spatial diversity is measured as the amount of accessible plots within a radius of three axial lines (left), and spatial integration and capacity overlaid in one map, showing the continuous variations of spatial capital (right) (Marcus, 2007)

and high diversity). Marcus's (2007) study shows that the accessible plots at the radius of three topological depths are mainly aggregated in the centre of the city where the plot sizes are smaller, and the network dense (Fig. 2.18 Left). By overlapping integration and accessible plots (Fig. 2.18 Right), varied spatial capitals were tentatively generated.

Moreover, apart from only using variables of urban form as attractions, there are studies (e.g. Legby et al., 2015; Marshhoodi and Pont, 2011; Stahle, 2016) capturing the accessibility to various urban amenities and resources to explore the relation between form and function. The correlation between the functioning of the city and its physical structure could contribute to proposing spatial interventions to improve socio-economic performance; forecasting the implications of changes to the spatial network structure in urban planning and design in direct or indirect relation to social cohesion and urban equality, such as reducing resident displacement and physical disruption in the regeneration projects (Karimi, 2012; Karimi et al., 2007; Stonor, 2017).

2.1.4 *The Spatial Analytical Approach*

The spatial analytical approach moves one step further than the configurational approach, and views the spatial structure and dynamics of cities as complex and emergent phenomena, in which there is an ambiguous relationship between social and economic interactions and the physical form of urban settlements (Kropf, 2009). Unlike modern urban planning which works backward, starting from the large-scale and ending in small-scale, spatial analytical tools rebuild urban growth process from the small-scale and extend it to the large-scale based on certain geometric rule (Helie, 2010). This approach is best characterized by the work of Michael Batty and the Centre for Advanced Spatial analysis, using a range of methods and models including GIS, cellular automata, entropy-based, agent-based models and fractals.

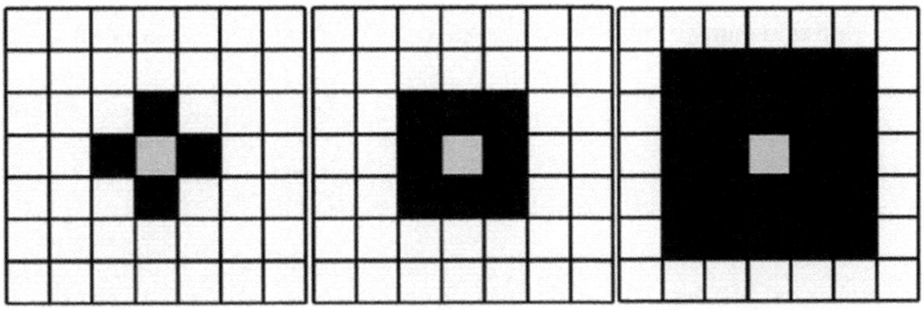

Fig. 2.19 Illustrations of the cell neighbourhood (Shiffman et al., 2012)

Among these tools, cellular automata model (CA) is a typical one. It was first introduced in 1948 by Neumann and Ulam to model complex dynamic systems. CA models could be classified into three categories according to their objective: descriptive models, which analyse the factors and dynamics that govern the evolution of urban land; predictive models, which simulate land use change in the near future; prescriptive models, aimed at optimizing configuration of land uses (Batty, 2007; Santé et al., 2010).

CA models are dynamic models composed of local interactions between cells on a regular dimensional grid. Each cell may be in a predetermined state. As simulation progresses, the state of each cell depends on its previous state and on the state of its neighbouring cells according to a set of transition rules. As Fig. 2.19 shows, the neighbouring cells are often defined as 4 cells surrounding the cell, known as Neumann neighbourhood, or 8 cells known as Moore neighbourhood, or even more, 24 cells.

In urban studies, cells in a CA model 'most appropriately' represent plots or parcels or their simple aggregations but might fall somewhere between parcels and census tracts or other administrative aggregations depending on the source of the data, such as (1) binary values (urban, non-urban); (2) different land uses; (3) population density or values of buildings and so on. The state of each cell is determined by its previous state and the state of its neighbouring cells according to a set of transitional rules. A cell space is composed of square cells of different resolutions (of 10 m to 1 km) (Santé et al., 2010). Take Stanilov et al.'s study (2011) for example, the cellular automata model is adopted to simulate the historical growth of west London, which has a long-time horizon from 1875 to 1915, relating to the evolution of urban environment (Fig. 2.20). The spatial relationship between land uses and physical environment (transportation infrastructure and the network of CBD remains consistent through time, which constitute a basic code for urban growth (Stanilov, 2012). The findings suggest that planning plays an important role in shaping the patterns of metropolitan expansion. In short, the CA model may not provide accurate or predictive descriptions but are able to 'strip the processes of city growth to their bare essentials, and thus to uncover the basic mechanism at work (Batty, 2007).

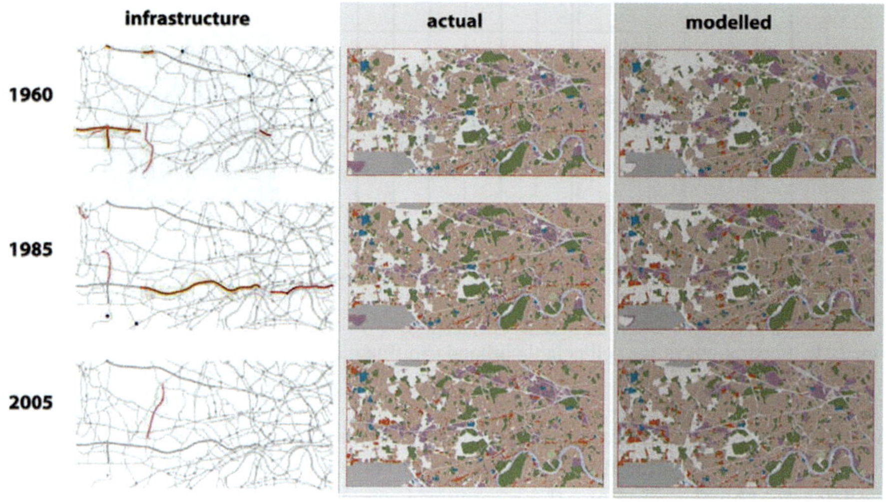

Fig. 2.20 Comparison of actual and modelled growth patterns (Stanilov, 2012)

This approach has effectively engaged in simulating spatial and temporal processes of urban growth in many urban studies, investigating the genetic rules which govern the urban growth, generating predictions of the future changes of spatial structure and assessing the impact of planning decisions (Batty & Xie, 1994; Batty et al., 1999; Wu & Webster, 1998; Kropf, 2007, 2018).

To sum up, each morphological approach focuses on slightly different aspects of urban form and uses different methods and tools. For example, the historico-geographical and process-typological approaches have a historical root on a qualitative basis while the configurational and spatial analytical approaches concern the relationship between form and use on a quantitative basis. In the framework of Gauthier and Gilliand (2006), different morphological approaches are classified and summarized as cognitive (descriptive) / normative (prescriptive) based on their epistemological perspectives. For instance, the historico-geographical approach is recognized as cognitive, describing urban form and its historical change, while the process-typological approach is categorized as normative, as it aims to produce new urban forms. The configurational approach is also normative, which allows the testing of different alternatives for transformation of street systems and provides future interventions. Yet this two-dimension is not that clearly cut. Recent developments in the historico-geographical approach (e.g. Gu & Xu, 2014; Whitehand & Gu, 2007a, 2007b), focus on its application in urban conservation and landscape management, which can be seen as normative contributions.

The above discussions have demonstrated that different morphological schools and associated methods and techniques across different disciplines could provide rigorous descriptions, explanations and prescriptions of the complex and diverse built environment (Larkham and Samuels, 2008; Whitehand, 2013). Because of its rigour and interdisciplinary character, urban morphology provides sound basis for studies into diverse aspects of environmental performance, microclimate, acoustics, movement, ecology and land use (Kropf, 2018). For designers and planners, morphological investigations provide significant insight into urban form as a medium for character assessment, heritage management, and urban design codes and guidance in recent years, contributing to the essential relationship between research and practice (Hall, 2008, 2013; Marshall & Çalişkan, 2011; Scheer, 2010; Whitehand, 2005, 2013).

2.2 Problems in Historical Urban Landscape Conservation and Management

During the rapid growth of cities in the past fifty years of urban development, great opportunities have been created to enhance the quality of life, access to services and transfer of technology. Yet at the same time, several problems have arisen including environmental stress, pressure on housing and urban services, social inequalities and so on (Bandarin & Van Oers, 2012). The historic inner cities especially in developing countries have become the focus of urban development and regeneration processes, experiencing great pressure for change (Bandarin & Van Oers, 2012).

Facts like this should have implications for what we think about urban conservation, but frequently urban planners have weak grounding in research on urban form (Whitehand, 2007). It is not uncommon for planning to focus on land use with ground plan and building form comparatively being less considered (Hall, 2008, 2013). Land use zoning which originated from the fourth *Congrès internationaux d'architecture moderne* (CIAM) has been perceived as a prevalent and efficient tool to plan the new or renewal projects. Yet it separates different uses and prevents a diverse urban community. Massive areas created are devoted to single uses (Christopher, 1966; Scheer, 2010). This is often accompanied with high speed of change in the shape of large and monolithic built forms with little or no reference to existing urban form and context. Land use zoning tends to be ineffective in guiding positive management of change to the historical urban landscape. These newly built forms destroy the intricate physical, functional and social microstructure that has taken so long to heal (Kropf, 2013).

This problem is partly due to the fact that the historic awareness of urban planners largely remains at the level of describing individual features. There is little sense of how they fit together to form an interrelated entity and are part of a process of change (Whitehand et al., 2011). And the absence of a systematic method of characterizing and delimiting the areas to be conserved and protected is widespread in urban conservation and urban landscape management. There are often ill-informed boundaries of a conservation area, cutting across the urban landscapes that are products of a city's historical development. These deficiencies further speed up the disappearance and fragmentation of historical urban landscapes (Whitehand, 2009).

Moreover, the fast-changing urban landscape has involved rapid economic development in competition for markets and investments, which in turn has had effects on the physical and socio-cultural aspects of cities. Much of the historic fabric has been altered by a top-down operation of aggressive real estate or theme-park developments. The local community has been removed from the place or driven out by high rents and cost of living. In view of this, researchers have long argued that historic districts could serve as a powerful catalyst for economic and social regeneration, where inhabitants should be able to conduct their lives, without which the essential ingredients of the spirit of place are lacking (Jacobs, 1961; Koh & Melic, 2017; Rodwell, 2008, 2009).

The notable effort is that UNESCO's recommendation on historical urban landscape has broadened the value of urban heritage to include the new use and enjoyment of the urban space that defines the city as 'living heritage' (UNESCO, 2011). It aims at preserving the quality of human environment, enhancing the productive and sustainable use of urban space while promoting social and economic diversity (Bandarin & Van Oers, 2012). It is within this context that the challenge of conserving and managing historical urban landscape has shifted to synergizing the heritage preservation and socio-economic development (Van Oers, 2006). In response to this challenge, a critical understanding of the transformation processes of the city's historical development as well as an integrated approach to determining limits of acceptable change are required (Van Oers, 2006, 2010).

Urban morphology has the potential to meet this challenge because of its rigorous description of the built environment that is also the medium of urban planning and design (Kropf, 2018; Marshall & Çalişkan, 2011; Rodwell, 2009). Urban morphology could provide enhanced understandings of the physical evolution and socio-economic performance of urban fabric. On the one hand, urban morphology views historicity as the key factors of urban landscapes. The analysis of the historico-geographical structure could strengthen the historical awareness among urban planners and designers and provide foundation for the recognition, delimitation and management of historic urban areas. On the other hand, urban morphology could analyse how the existing socio-economic activities are organized in the built environment, and identify what the potentials are while proposing for management of change to improve the quality of life and urban space.

Moreover, to explore the synergies between conservation and socio-economic development to improve the sustainability of planning and design requires the integrated efforts of different morphological approaches, as each of them has a certain focus on the built environment in the fields of architecture, geography, history and urbanism. Take the historico-geographical approach for example, recognizing plan units has proven not easy to convince academics in other disciplines that valuable data are embedded in urban landscapes (Larkham, 2006), neither could the boundary delimitation of plan unit be precisely replicated by researchers (Whitehand, 2009). This may indicate that additional rigorous and sound evidence is necessary to enhance plan unit recognition. For the configurational approach, a full interpretation of the extent to which these spatial analyses are socio-spatially meaningful can hardly be achieved without thorough historical research (Griffith et al., 2012). The challenge raised by this synergy turns out to be how to select between different views of morphological traditions, and combine and coordinate them to contribute to the understanding and management of historical urban landscape in a period of unprecedented change.

References

Al-Sayed, K., Turner, A., Hillier, B., Iida, S., & Penn, A. (2014). *Space syntax methodology. A teaching textbook for the MSc spatial design: Architecture & cities.* Retrieved from https://pdf coffee.com/qdownload/alsayed-turner-et-al-2014-space-syntax-methodology-pdf-free.html.

Baker, N. (2010). A characterisation of the historic townscape of Central Hereford. *Hereford: Herefordshire County Council. Herefordshire Archaeological Report, 266.*

Baker, N., & Slater, T. (1992). Morphological regions in English medieval towns. In *Urban landscapes: International perspectives* (pp. 43–68).

Bandarin, F., & Van Oers, R. (2012). *The historic urban landscape: Managing heritage in an urban century.* Wiley.

Batty, M. (2004). *A new theory of space syntax.* Centre for Advanced Spatial Analysis (UCL).

Batty, M. (2007). *Cities and complexity: Understanding cities with cellular automata, agent-based models, and fractals.* The MIT press.

Bobkova, E., Marcus, L., & Pont, B. (2017). *Multivariable measures of plot systems: Describing the potential link between urban diversity and spatial form based on.* Paper presented at the 11th Space Syntax Symposium, Lisbon.

Caniggia, G., & Maffei, G. L. (1982). *Composizione architettonica e tipologia edilizia: Lettura dell'edilizia di base. 1.* Marsilio.

Caniggia, G., & Maffei, G. L. (2001). *Architectural composition and building typology: Interpreting basic building* (Vol. 176). Alinea Editrice.

Cataldi, G., Maffei, G. L., & Vaccaro, P. (2002). Saverio Muratori and the Italian school of planning typology. *Urban Morphology, 6*(1), 3–14.

Charalambous, N., & Mavridou, M. (2012). *Space Syntax: Spatial Integration Accessibility and Angular Segment Analysis by Metric Distance (ASAMeD).* Retrieved from Nicosia, Cyprus: https://www.accessibilityplanning.eu/uploads/pdf/3-ASAMeD-R.pdf.

Christopher, A. (1966). A city is not a tree. *Design Magazine, 206*, 46–55.

Conzen, M. R. G. (1960). Alnwick, Northumberland: A study in town-plan analysis. *Transactions and Papers (Institute of British Geographers)*, iii-122.

Conzen, M. R. G. (1964). *The urban morphology of Central Newcastle.* Paper presented at the the XXth International Geographical Congress 1964, Symposium in Urban Geography S2 Edinburgh, Scotland.

Conzen, M. R. G. (1988). Morphogenesis, morphological regions and secular human agency in the historic townscape, as exemplified by Ludlow. In *Urban historical geography: recent progress in Britain and Germany* (pp. 253–272).

Conzen, M. R. G. (2004). Morphogenesis and structure of the historic townscape in Britain. In M. R. G. Conzen & M. P. Conzen (Eds.), *Thinking about urban form: Papers on urban morphology 1932–1998.*

Crucitti, P., Latora, V., & Porta, S. (2006). Centrality in networks of urban streets. *Chaos: An Interdisciplinary Journal of Nonlinear Science, 16*(1), 015113.

Dhanani, A., Vaughan, L., Ellul, C., & Griffiths, S. (2012). *From the axial line to the walked line: Evaluating the utility of commercial and user-generated street network datasets in space syntax analysis.* Paper presented at the Eighth International Space Syntax Symposium.

Gauthier, P., & Gilliland, J. (2006). Mapping urban morphology: A classification scheme for interpreting contributions to the study of urban form. *Urban Morphology, 10*(1), 41.

Griffith, S., Greene, M., Reyes, J., & Castro, A. (2012). *The use of space syntax in historical research: Current practice and future possibilities.* Paper presented at the Conference Proceedings of the Eighth International Space Syntax Symposium.
Gu, K. (2001). Urban morphology of China in the post-socialist age: Towards a framework for analysis. *Urban Design International, 6*(3–4), 125–142.
Gu, K., Tian, Y., Whitehand, J., & Whitehand, S. M. (2008). Residential building types as an evolutionary process: The Guangzhou area, China. *Urban Morphology, 12*(2), 97–115.
Gu, K., & Xu, Z. (2014). Applying Conzenian and Caniggian ideas in China: Recent research advances and problems. *U+D Urbanform and Design (1)*, 10–21.
Hall, T. (2008). The form-based development plan: Bridging the gap between theory and practice in urban morphology. *Urban Morphology, 12*(2), 77–95.
Hall, T. (2013). The potential influence of urban morphology on planning practice. *Urban Morphology, 17*(1), 54–55.
Hansen, W. G. (1959). How accessibility shapes land use. *Journal of the American Institute of Planners, 25*(2), 73–76.
Helie, M. (2010). *Emergent urbanism-the role of urban complexity in the practice of urbanism.* Retrieved from http://emergenturbanism.com/urban-complexity-in-the-practice-of-urbanism/.
Hillier, Yang, T., & Turner, A. (2012). Normalising least angle choice in depthmap-and how it opens up new perspectives on the global and local analysis of city space. *Journal of Space Syntax, 3*(2), 155–193.
Hillier, B. (1996). Cities as movement economies. *Urban Design, 1*(1), 41–41.
Hillier, B., & Hanson, J. (1984). *The social logic of space.* Cambridge University Press.
Hillier, B., & Iida, S. (2005). Network and psychological effects in urban movement. In A. Cohn & D. Mark (Eds.), *Spatial information theory* (Vol. 3693, pp. 475–490). Springer.
Hillier, B., & Penn, A. (2004). Rejoinder to carlo ratti. *Environment and Planning B: Planning and Design, 31*(4), 501–511.
Hillier, B., Penn, A., Hanson, J., Grajewski, T., & Xu, J. (1993). Natural movement: Or, configuration and attraction in urban pedestrian movement. *Environment and Planning B: Planning and Design, 20*(1), 29–66.
Hillier, B., Turner, A., Yang, T., & Park, H. T. (2007). *Metric and topo-geometric properties of urban street networks.* Paper presented at the Proceedings Space Syntax. 6th International Space Syntax Symposium, Istanbul.
Hillier, B., & Vaughan, L. (2007). The city as one thing. *Progress in Planning, 67* (3), 205–230
Jacobs, J. (1961). *The death and life of great American cities.* Wiley.
Karimi, K. (1997). *The spatial logic of organic cities in Iran and the United Kingdom.* In Proceedings of Space Syntax 1st International Symposium London (pp. 1–17).
Karimi, K. (1999). *Urban conservation and spatial transformations.* Paper presented at the The Space Syntax Second International Symposium Proceedings, Brasillia.
Karimi, K. (2012). A configurational approach to analytical urban design: 'Space syntax' methodology. *Urban Design International, 17*(4), 297–318.
Karimi, K., Amir, A., & Shaiei, K. (2007). *Evidence-based spatial intervention for regeneration of informal settlements: The case of Jeddah central unplanned areas.* Paper presented at the 6th International Space Syntax Symposium, Istanbul.
Koh, M., & Melic, K. (2017). *A historic heart how heritage districts can make cities more liveable.* Retrieved from Singapore. https://www.clc.gov.sg/docs/default-source/reports/a-historic-heart.pdf.
Kropf, K. (1996). An alternative approach to zoning in France: Typology, historical character and development control. *European Planning Studies, 4*(6), 717–737.
Kropf, K. (2005). The handling characteristics of urban form. *Urban Design* (93), 17–18.
Kropf, K. (2009). Aspects of urban form. *Urban Morphology, 13*(2), 105–120.
Kropf, K. (2011). Morphological Investigations: Cutting into the substance of urban form. *Built Environment, 37*(4), 393–408. https://doi.org/10.2148/benv.37.4.393

Kropf, K. (2013). Evolution and urban form: Staking out the ground for a more mature theory. *Urban Morphology, 17*(2), 136–149.
Kropf, K. (2017). *Bridging configurational and urban tissue analysis.* Paper presented at the Proceedings, 11th International Space Syntax Symposium, Lisbon.
Kropf, K. (2018). *The handbook of urban morphology*. Wiley.
Larkham, P. J. (1996). *Conservation and the city*. Taylor & Francis.
Larkham, P. J. (2006). The study of urban form in Great Britain. *Urban Morphology, 10*(2), 117–141.
Legby, A., Pont, M. B., & Marcus, L. (2015). Street interaction and social inclusion. In *Suburban urbanites: Suburbs and the life of the high street* (pp. 239–262).
Li, Y., & Gauthier, P. (2014). The evolution of residential buildings and urban tissues in Guangzhou, China: Morphological and typological perspectives. *Urban Morphology, 18*(2).
Lilley, K. D. (1998). Taking measures across the medieval landscape: Aspects of urban design before the renaissance. *Urban Morphology, 2*(2), 82–92.
Lilley, K. D. (2000). Mapping the medieval city: Plan analysis and urban history. *Urban History, 27*(1), 5–30.
Marcus, L. (2005). *Plot syntax: A configurational approach to urban diversity.* Paper presented at the Proceedings, Fifth international space syntax symposium.
Marcus, L. (2007). *Spatial capital and how to measure it: An outline of an analytical theory of the social performativity of urban form.* Paper presented at the Proceedings, 6th International Space Syntax Symposium, Istanbul.
Marcus, L. (2010). Spatial capital: A proposal for an extension of space syntax into a more general urban morphology. *The Journal of Space Syntax, 1*, 30–40.
Marcus, L., Pont, M. B., & Bobkova, E. (2017). *Cities as accessible densities and diversities.* Paper presented at the 11th Space Syntax Symposium.
Marshall, S. (2009). *Cities design and evolution.* Routledge.
Marshall, S., & Çalişkan, O. (2011). A joint framework for urban morphology and design. *Built Environment, 37*(4), 409–426.
Martin, L., & March, L. (1972). *Urban space and structures* (Vol. 1). Cambridge University Press.
Mashhoodi, B., & Berghauser Pont, M. (2011). *Studying land-use distribution and mixed-use patterns in relation to density, accessibility and urban form.* Paper presented at the International Seminar on Urban Form, Montreal.
Moudon, A. V. (1997). Urban morphology as an emerging interdisciplinary field. *Urban Morphology, 1*(1), 3–10.
Muratori, S. (1959). *Studi per una operante storia urbana di venezia.* Instituto poligrafico dello Stato.
Palaiologou, G., & Griffiths, S. (2019). The uses of space syntax historical research for policy development in heritage urbanism. In M. Obad Šćitaroci, B. Bojanić Obad Šćitaroci (Eds.), *Cultural urban heritage* (pp. 19–34). Springer.
Ratti, C. (2004). Space syntax: Some inconsistencies. *Environment and Planning B: Planning and Design, 31*(4), 487–499.
Rodwell, D. (2008). Urban regeneration and the management of change: Liverpool and the historic urban landscape. *Journal of Architectural Conservation, 14*(2), 83–106.
Rodwell, D. (2009). Urban morphology, historic urban landscapes and the management of historic cities. *Urban Morphology, 13*(1), 78.
Samuels, I. (2008). Typomorphology and urban design practice. *Urban Morphology, 1*(12), 58–62.
Santé, I., García, A. M., Miranda, D., & Crecente, R. (2010). Cellular automata models for the simulation of real-world urban processes: A review and analysis. *Landscape and Urban Planning, 96*(2), 108–122.
Scheer, B. (2010). *The evolution of urban form: Typology for planners and architects.* Routledge.
Shiffman, D., Fry, S., & Marsh, Z. (2012). *The nature of code: Simulating natural systems with processing.* D. Shiffman.

Slater, T. R. (1990). English medieval new towns with composite plans: Evidence from the Midlands. In T. R. Slater (Ed.), *The built form of western cities: Essays for MRG Conzen on the occasion of his eightieth birthday*. Leicester University Press.
Stahle, A. (2016). *Closer together: This is the future of cities*. Dokument Press.
Stahle, A., Marcus, L., & Karlström, A. (2005). *Place syntax: Geographic accessibility with axial lines in GIS*. Paper presented at the ffth international space syntax symposium, Delft.
Stanilov, K. (2012). Planning the growth of a Metropolis: Factors influencing development patterns in West London, 1875–2005. *Journal of Planning History, 00*, 1–21.
Stanilov, K., & Batty, M. (2011). Exploring the historical determinants of urban growth patterns through cellular automata. *Transactions in GIS, 15*(3), 253–271.
Stonor, T. (2017). A human-focused "science of building and cities" in the context of new techniques and new data. *Time+Architecture,* (5), 63–66.
Thomas, R. M. (2006). Mapping the towns: English Heritage's urban survey and characterisation programme. *Landscapes, 7*(1), 68–92.
Thomas, R. M. (2018). Conservation, heritage and urban morphology-further thoughts. *Urban Morphology, 1*(22), 71–73.
UNESCO. (2011). *Recommendation on the historic urban landscape*. Paper presented at the Records of the General Conference 36th Session.
van Nes, A., & Yamu, C. (2021). *An introduction to space syntax in urban studies*. Springer.
Van Oers, R. (2006). Preventing the goose with the golden eggs from catching bird flu–UNESCO's efforts in safeguarding the historic urban landscape. *Cities between Integration and Disintegration: Opportunities and Challenges, ISoCaRP Review, 2*.
van Oers, R. (2010). Managing cities and the historic urban landscape initiative–an introduction. In *Managing historic cities* (pp. 7–17).
Whitehand, J., & Gu, K. (2007a). Extending the compass of plan analysis: A Chinese exploration. *Urban Morphology, 11*(2), 91–109.
Whitehand, J., Gu, K., Whitehand, S. M., & Zhang, J. (2011). Urban morphology and conservation in China. *Cities, 28*(2), 171–185.
Whitehand, J. W. R. (1977). The basis for an historico-geographical theory of urban form. *Transactions and Papers (Institute of British Geographers), 2*(3).
Whitehand, J. W. R. (2005). Urban morphology, urban landscape management and fringe belts. *Urban Design, 93*, 19–21.
Whitehand, J. W. R. (2009). The structure of urban landscapes: Strengthening research and practice. *Urban Morphology, 13*(1), 5–27.
Whitehand, J. W. R. (2013). Urban morphological research and practice. *Urban Morphology, 1*(17), 3–4.
Whitehand, J. W. R. (2015). Conservation, heritage and urban morphology. *Urban Morphology, 19*(2), 115–116.
Whitehand, J. W. R., & Gu, K. (2007b). Urban conservation in China: Historical development, current practice and morphological approach. *Town Planning Review, 78*(5), 643–670.
Whitehand, J. W. R., Gu, K., & Conzen, M. P. (2014). The typological process and the morphological period: A cross-cultural assessment. *Environment and Planning B: Planning and Design, 41*(3), 512–533.
Yang, T., & Hillier, B. (2007). *The fuzzy boundary: The spatial definition of urban areas*. Paper presented at the 6th International Space Syntax Symposium, Istanbul.

Chapter 3
Traditions of Urban Morphology: Opportunities and Challenges

Abstract Each morphological approach has its strength and weakness in description, analysis and prescription of urban form. Integrating different morphological approaches—the historico-geographical, process-typological, configurational and spatial analytical approach—could provide enhanced understandings of urban form, which is one of the major challenges facing urban morphology. Synthesis of the first two approaches has been extensively studied. In recent years, there have been notable attempts at conceptualizing the combination of the historico-geographical and configurational approaches in the study of historical transformation and socio-economic performance of urban form. Yet to what extent the configurational approach could inform plan unit division and measure socio-economic performance of morphological region (landscape unit) remains not fully clear. Further investigations are needed to explicate to what extent and manner in which way there two approaches can be combined, and how the combination can contribute to enhancing our understandings of urban form. Moreover, the relation between morphological research and practice has long been rather weak. How the enhanced understanding of urban form could inform planning and management of historical urban landscape awaits further exploration.

Keywords Urban morphology · Historical transformation · Socio-economic profile · Urban form

There have been excellent attempts to summarize the theories of different morphological approaches. Moudon (1997) has identified three shared theoretical bases: 'firstly, the city or town can be "read and analysed" via its physical form, which is defined by three fundamental urban elements: buildings, plots and streets; secondly, urban form can be understood at different levels of resolution, such as building/plot, street/block, the city and the region; thirdly, urban form can be understood historically since the elements undergo continuous transformation'. Thus form, resolution and time are important components of urban morphological research. Gauthier and Gilliland (2006) provide another framework for classifying different schools of thought based on Moudon (1994) and Levy (2005)'s distinctions. There are four quadrants identified: descriptive (cognitive), prescriptive (normative), externalist and internalist (Fig. 3.1).

X. Li and Y. Zhang, *Conserving and Managing Historical Urban Landscape*, The Urban Book Series, https://doi.org/10.1007/978-981-19-4222-8_3

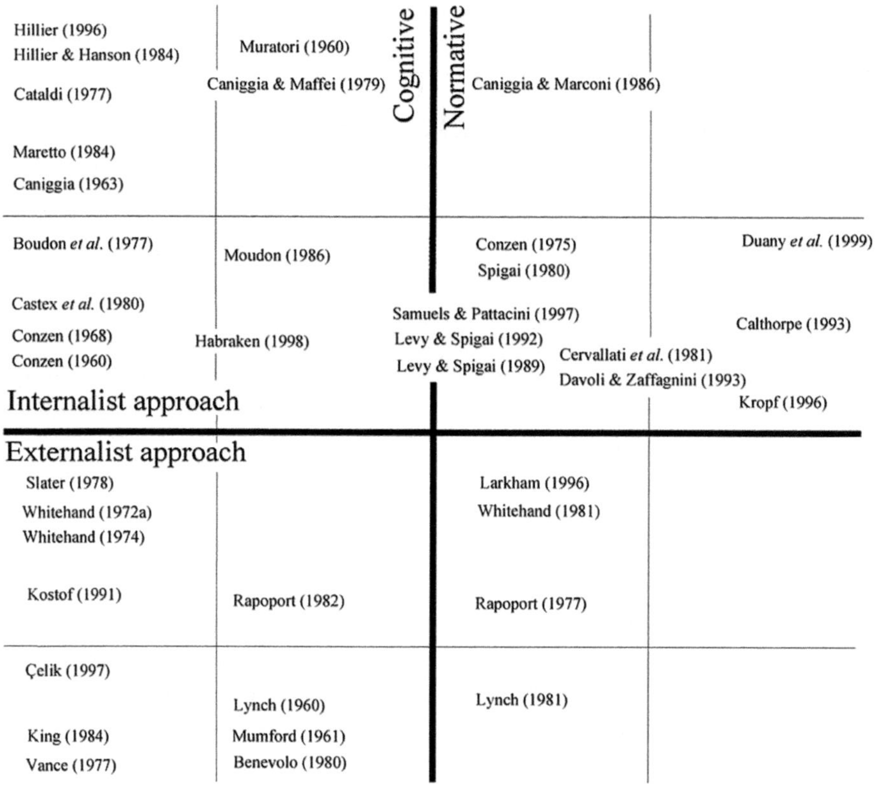

Fig. 3.1 Classification of morphological approaches (Gauthier & Gilliland, 2006)

Urban morphology could provide systematic description, explanation and prescription of the formation and transformation process of urban form. Its major concern is on the one hand with understanding the processes that have produced what is there in the built environment and how and why that changes, on the other hand with providing insights on prescription for the transformation or conservation of existing forms. These two aspects are summarized as descriptive and prescriptive (Gauthier & Gilliland, 2006; Moudon, 1997). The descriptive aspect is concerned with how cities are built and why. This is the primary focus of the historico-geographical approach that concerns successes and failures of past societies in their attempts to shape the built environment. There is also the focus of the configurational approach which explores the underlying relationship between form, movement and use (Hillier, 1996, 1999). The prescriptive aspect concentrates how cities should be built. It mainly includes studies of the configurational approach (Karimi, 1999, 2012), while some historico-geographical studies (Whitehand & Gu, 2007; Whitehand et al., 2011) in the heritage conservation and management also belong to this category.

Urban morphology could also define and illustrate the principles of change in urban landscapes. On the one hand, it explains how different agents change the

Table 3.1 Features of different approaches (Kropf, 2009)

Morphological approaches	Features	Relations between features
Historico-geographical	Area or patch	Patch hierarchy
Process-typological	Module	Modular hierarchy
Configurational	Line	Network hierarchy
Spatial analytical	Cluster	Aggregate cluster

physical form over time. For example, people design, demolish buildings and areas, promote or control development through planning regulations and local organization. On the other hand, it uncovers the internal logic for a city's physical formation and transformation. These two aspects are summarized by Gauthier and Gilliland (2006) as externalist and internalist. The externalist deals with the transformation of the built environment, but which rely mainly on the external conditions of development. It includes studies using the historico-geographical approach to explore the range of agents active for conservation and planning together with the effects of their characteristic on the urban landscape (Whitehand & Gu, 2010). The internalist conceptualizes the built environment as a dynamic system with relative autonomy. It includes studies (Conzen, 1960, 1964) using the historico-geographical approach and studies (Hillier, 1996, 1999; Hillier & Hanson, 1984) using the configurational approach.

Due to the different disciplinary roots and theoretical propositions, combining and coordinating different schools of thought and approaches to enhance the description, explanation and prescription of urban form is one of the key challenges facing urban morphology (Whitehand, 2012, 2015, 2017). Kropf (2009, 2011, 2014) seeks the common ground between different approaches by using urban form as a common reference and zooming into the form elements and patterns. The same set of form elements (building, plot and street) of urban form is used to identify the different features captured by different morphological approaches (Table 3.1). These different features and structures are not mutually exclusive, rather, they are related to one another. Their relationship and the hierarchical structure of form elements are further explained by Kropf (2014) using a multi-level diagram based on the relationship of part to whole.

This multi-level generic diagram was generated by revisiting the basic concept in the historico-geographical, process-typological and configurational approaches (Fig. 3.2). It visualizes the pattern of relationships in both vertical and horizontal perspective. Vertically, plot is made up of the building and its associated open space. Plot series is an aggregate of similar plots, lined up along the route which provides access and movement. The street or simple tissue is a combination of buildings, streets and plots, similar to Conzen's definition of plan unit. The more complex combinations of simple tissues are called urban tissue. This part-to-whole hierarchy has always been the object of the historico-geographical and process-typological approaches. Horizontally, the street space is an essential composition of structures, providing access and movement to plots, plot series, simple tissue and urban tissue. It

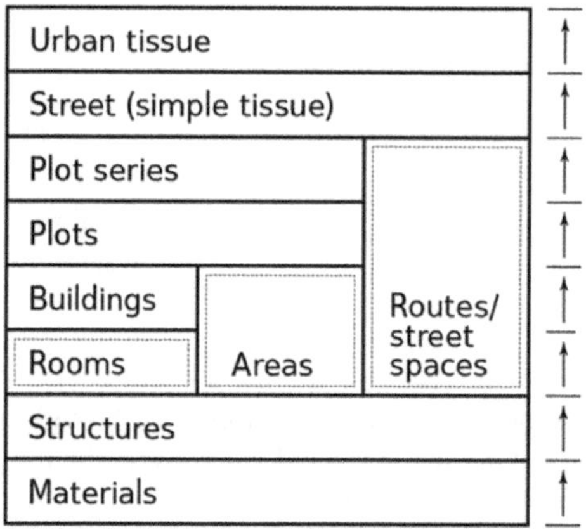

Fig. 3.2 Multi-level diagram showing vertical and horizontal perspective of form (Kropf, 2014)

is also a focus of the configurational approach. From this analysis of the compositional hierarchy, it could be argued that these morphological approaches are complementary and could work together to provide a broader view of built form.

Consistent with Kropf's principle of form elements, Oliveira et al. (2015) continued to explore the bridge between four morphological approaches: the historico-geographical approach, the process typological approach, the configurational approach and the spatial analytical approach. It is argued that the concept of morphological region (landscape unit) has the potential to provide a framework to coordinate different approaches through a case study in Porto.

However, these important works have limitations: the generic structure developed by Kropf lacked illustrations of how it could improve practical understandings, while Oliveira's work did not go further in the establishment of an integrated view. As a result, the main goal of combining and coordinating different morphological schools is still unfulfilled (Whitehand, 2019). Out of the four morphological approaches, explorations on combining and coordinating two or three morphological schools have a significant development.

3.1 Urban Morphology and the Historical Transformation of Urban Form

Combining the historico-geographical and process-typological approaches has been extensively studied due to their similar historical roots (Chen & Thwaites, 2013; Kropf, 1993; Maffei & Whitehand, 2001; Li & Gauthier, 2014; Whitehand et al., 2014). The term typo-morphology was first coined by Italian architect Carlo Aymonino in the 1960s to reflect the dialectic between building typology and urban morphology (Aymonino et al., 1966). Kropf (1993) started to integrate morphology theory with typology theory by exploring the relationship between morphological region and urban tissue. Though the definition of 'typomorphology' is still absent, it has already been suggested as a combination of both approaches.

The historico-geographical approach concentrates how constellations of forms develop in different periods, there has been little discussion on how forms created in one morphological period are succeeded by those created in the next where by period types arise. The typological process is very much concerned with this (Maffei & Whitehand, 2001). Thus the combined framework could help to understand both physical structure of individual buildings and historical development of urban areas as well as their transformations at different levels of resolution. Moreover, thinking about both approaches opens up questions of how proposals might be formulated about the future of urban forms. It has been applied and tested in different urban contexts, providing both in-depth understanding of the identity of places and suggestions for urban designers and policy makers to produce culturally sustainable urban forms.

However, establishing a dialogue between the historico-geographical and configurational approaches has been proved not easy and straightforward. This has been largely hindered by their different disciplinary traditions and theoretical propositions: the primary purpose of the historico-geographical approach is to describe and explain the historical development of cities through historical maps, archival studies and site survey; while the primary focus of the configurational approach is to inform future design interventions based on the mathematical and quantitative investigations of urban form.

Larkham (2006), among others, argued that there exists a great opportunity for exploring the complementarity of the historico-geographical approach and the configurational approach. It is suggested that both approaches share the same focus of investigation—the structure of space embodies and explains its socio-economic and cultural processes (Griffiths et al., 2010), and the common object of study—the street—is an essential built form element for both delimiting the morphological regions (plan units) and recognising accessibility (Kropf, 2017; Oliveira et al., 2015) (Fig. 3.2).

On this basis, significant efforts have been made to combine these two approaches to better elaborate the historical process of urban form (Griffiths et al., 2010; Kropf, 2017; Oliveira et al., 2015; Pinho & Oliveira, 2009). Kropf (2017) study has found that on the one hand, the configurational approach (route structure analysis) could provide a basic profile of route types within a system that serves as rigorous foundation for identifying urban tissue (plan units); on the other hand, the historico-geographical approach is able to articulate the place-specific context for the configurational analysis (Fig. 3.3). The findings suggest that combining historico-geographical and configurational approaches has the potential to yield additional information that helps to improve our understandings of urban form: such as the relationship between different tissues; the relationship of the international parts to the whole; the relationships between a given tissue and the settlement as a whole. Kropf's work is a first step in the identification of urban tissue from a composite view. While route structure focuses on configurational types, there are other analytical methods that focus on measures—space syntax and place syntax. The quantitative measures like integration

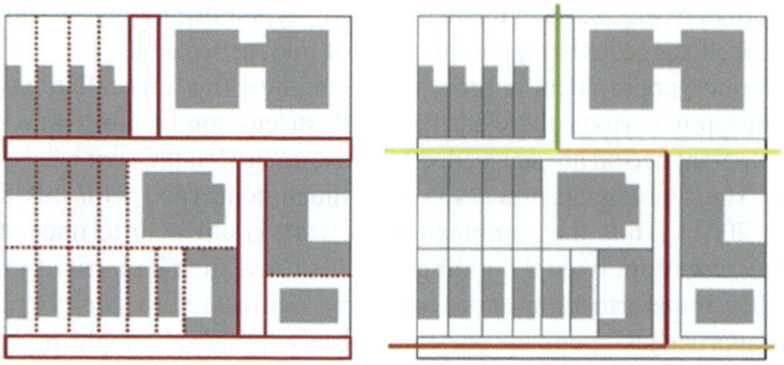

Fig. 3.3 Street as the common research object in both the historico-geographical and configurational approaches. Right: street space in plan units. Left: segment lines representing the street network

and connectivity could yield additional inference for delimitation of character areas such as plan units, which awaits further exploration.

Yet existing studies of combining the historico-geographical and configurational approaches primarily revolve around how the place-specific knowledge about the change and continuity of urban form can explain its generic configurational properties of the street network, whereas investigations on how the configurational types and measures of the street network could inform identification of character areas such as plan units are only in its infancy (Zhang & Li, 2020). Thus, our knowledge as to what new insights that the combination of the historico-geographical and configurational approaches could offer remains not fully clear.

Further investigations are needed to explicate to what extent and in which way these two approaches can be combined, and how the combination can contribute to enhancing our understandings of urban form. Another neglected opportunity is that combining the two approaches could shed light on the transformation process of urban form. Through systematically examining and comparing urban form across different morphological periods, the generative rules of street configuration[1] can help explain urban growth and development, in particular how different plan units aggregate to form the entire settlement. The historico-geographical approach may provide a way to examine the role of street configuration played in the formal changes of a city within a specific urban context. Therefore, further investigations are needed to explicate to what extent and manner in which way there two approaches can be combined, and how the combination can contribute to enhancing our understandings of urban form.

[1] The generative rules of urban space refer to the structure of street network and its relation with urban functions: cities comprise a small number of longer lines with near straight connections that seek to maximize movement driven by microeconomic factors, and a large number of shorter lines with near right-angle connections that are configured to restrain and structure movement driven by social and cultural factors (Hillier, 1996, 2007).

3.2 Urban Morphology and the Socio-Economic Profile of Urban Form

Seeing the built environment in terms of its historic value is fundamentally limited in its ability to see other dimensions. It can't touch the issues driving people's day-to-day lives (Kropf, 2009). Jacobs (1961) was the first to explore human activity in the built environment that contributes to urban quality (Montgomery, 1998). She identifies four essential determinants that influence human activity: a mixture of primary use, intensity, permeability of urban form and a mixture of building types, ages and sizes. Gehl (1989) argues that successful urban places are based on street life, and activity occurs in buildings and places. Pedestrian flow is a prerequisite of the presence of an active street life. Lynch (1984) put forward a few dimensions of urban quality. Vitality is defined as the degree to which the form of settlement supports the vital functions. Access is defined as the quantity and diversity of persons, activities, services that could be reached, and its reach and distribution area are a basic index of urban quality. Fit is defined as the degree to which the form and capacity space match the pattern and quantity of human activity. These studies have considered form with respect to use through a number of quality-of-life parameters. These parameters are not mutually exclusive; in fact, they are interrelated in many ways. For example, mixed use and easy access encourage pedestrian flow and provide an opportunity for hustle and bustle urban life. This socio-spatial relationship should be more explicit regarding its descriptions and explanations of the physical dimensions and their implication on social, cultural and economic conditions, as well as its proposals for improving urban life in the future (Bosselmann, 2008). For this purpose, urban morphology could make a contribution, each morphological approach elaborating a different aspect of form.

For the historico-geographical and configurational approaches, the study of physical form and use is also the common research object of both approaches (Kropf, 2009). Land use has an impact on the socio-economic profile of urban areas. In the historico-geographical approach, the pattern of land use and building utilization presents another point of reference for distinguishing landscape units[2] and understanding the human-built form relation. The configurational approach (space syntax) also includes the form–function relationship by exploring the impact of the relationship between street configuration and movement on land use patterns. Yet how the street system relates to land use is a relatively unexplored area in the historico-geographical approach where space syntax can make a contribution.

Within the field of urban morphology, the unit of land use is plot that is one of the fundamental form elements (Fig. 3.4). Plot is also an area of land, usually a unit of ownership. Thus, plot is defined in terms of both land use and physical form (Kropf, 2009, 2018), and the attributes of the plot reflect not only a time period of history,

[2] Landscape unit is defined as areas having a degree of unity in their urban form derived from the combination of ground plan (street, plot and building), building form and land and building utilization (Conzen, 2004).

but also the socio-economic conditions at the time of land development and building (Moudon, 1997).

With this in mind, the past decade has witnessed some efforts to combine the historico-geographical and configurational approaches to understand the 'spatial milieu of human social life'. Griffith et al. (2010) study has traced existing forms back to different historical periods, seeking to present a complex picture of the suburban built forms and particularly of the distribution of socio-economic activity in the Greater London suburbs. It is found that the continuity in the socio-economic life of suburban centres is related to the persistence of historical road networks and the relative positions of the street plan of a particular settlement in a larger regional road network (Fig. 3.5). It is concluded that space syntax complements Conzenian analysis by measuring and identifying the relative locations of street plans and its implications on socio-economic activities, while the historical research by the Conzenian analysis provides a thorough interpretation of the generic properties of street network. Griffith's work is a first step of combining the historico-geographical and configurational approaches in providing morphological explanations for the long-term socio-economic activity. This has tremendous implications for both the planning and management of urban areas in the future.

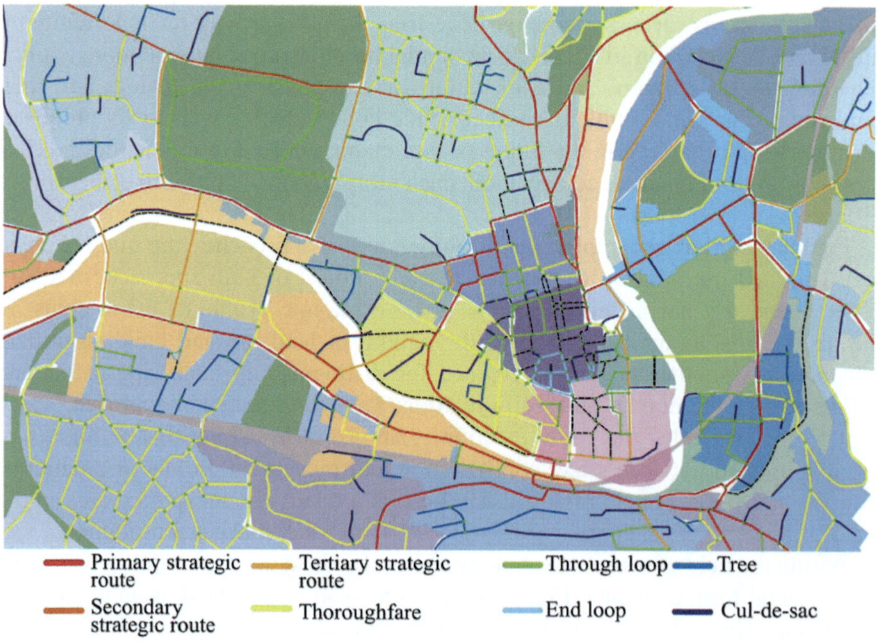

Fig. 3.4 Route structure types and tissue types in Central Bath. Colour-coded route structure identifies the different types and subsystems of routes. Coloured areas represent the central and fringe urban tissues. The patterns of routes are one of the key characteristics that define tissues. Putting the two drawings together highlights the distinct patterns in each tissue. For example, there is clearly a higher density of routes within the center, associated with more active, mixed uses (Kropf, 2017)

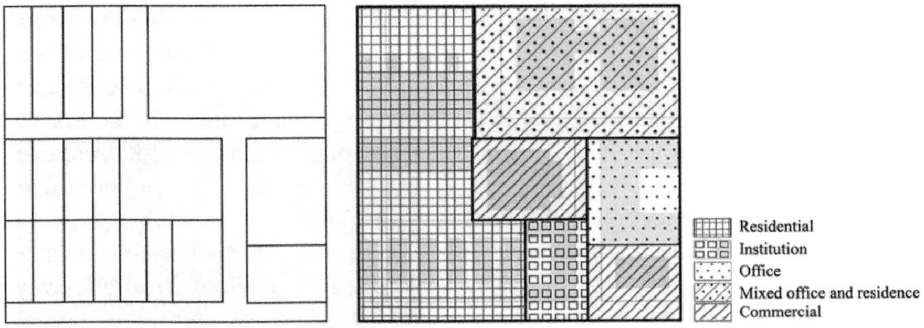

Fig. 3.5 Right: the plot defined as the one of the fundamental form elements in urban morphology. Left: plot as a unit of land use that are residential, institutional, office and so on

Over the recent years, another issue in urban morphology is the relation between research and practice. Monteiro and Pinho (2021) attempt to address this challenge by integrating the historico-geographical, process-typological and configurational approaches. It aims to provide a point of reference for future planning and development proposals based on a systematic analysis of built form. The configurational approach is adopted to test and simulate the proposed new streets and examine its impact on the accessibility of the morphological unit and the city (Fig. 3.6). The design interventions of new streets aim at increasing accessibility of morphological units and their urbanity,[3] based on the relationship between street network structure and land use that may potentially contribute to the socio-economic performance of urban areas. A morphological zoning plan is proposed along with regulation which provide the basis for rules of conservation, transformation and expansion. Monteiro and Pinho's work is a contribution to the growing body of research that bring together different morphological traditions to understand the complex urban form and inform planning decisions (Fig. 3.7).

Yet in space syntax, the socio-economic activities are treated as an outcome to substantiate the results of street configuration rather than as an equally important urban element. In reality, people's actual experience of space is related to not only street configuration, but also attractions such as land use (Stahle et al., 2005). As introduced in Sect. 2.1.3, new emerging analytical methods in the configurational approach like Place Syntax include attraction into the analysis. For example, functional data like land use or Point of Interest data (POI) can be linked to plots or actual building forms (Fig. 3.8). A land use zone identified as residential might contain other uses such as office and small-scale retail. Place syntax could identify the current activities accommodated by building or plot, irrespective of its original use and building type such as 'housing' or 'office'. In this way, it allows for a more accurate and finer-scale analysis of the form–function relationship.

---

[3] Urbanity is exposed in the relationship between the social and spatial dimensions, between people and the physical structure (Holanda, 2011).

Fig. 3.6 Left: continuity in development of plots in Surbiton from c. 1820. The thick-grey lines represent roads from c. 1820 where the same routes are still extant in contemporary Surbiton. Clusters of plots dating from c.1820 are found adjacent to routes dating from the same period. Right: potential through movement at radius-400 m showing non-residential land uses in Surbiton c. 2005. A significant number of non-residential uses are evident on Surbiton Road and Ewell Road which are locally accessible streets (Griffith et al., 2010)

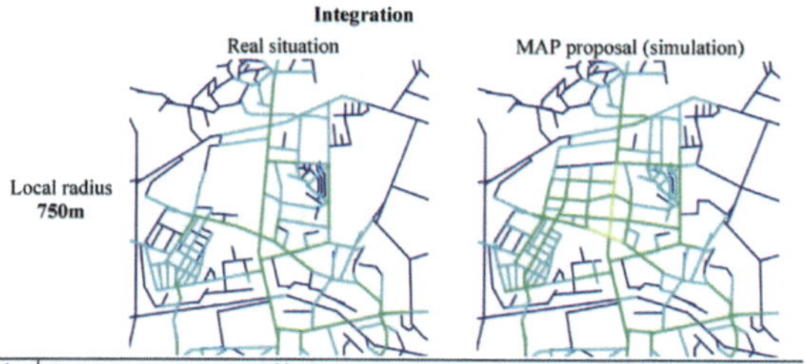

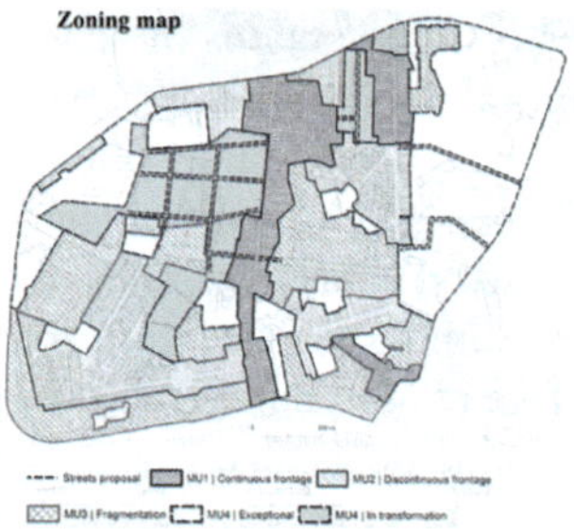

		Real situation	MAP proposal (simulation)
Study area	Average	80	90
	Variation	–	+ 13%
Rua do Amial	Average	129	153
	Variation	–	+ 18%

Fig. 3.7 Left: angular segment analysis. Comparative of existing street systems and that proposed. Right: zoning map. Five morphological units are identified along with transformation rules for existing urban forms and expansion rules for new urban forms (Monteiro & Pinho, 2021)

This has opened up great opportunities for measuring and analysing urban form in relation to the socio-economic forces that shape it across different urban scales, such as building, plot, plot series and plan unit (landscape unit). Besides, plot/building-based GIS offers the empirical data that could address contemporary urban areas, moving the centre of urban morphological research from the applications in urban conservation to management of future urban development. Thus, combining the historico-geographical and configurational analysis has the potential to provide sound bases for future transformation of urban form. Studies along this direction await further explorations.

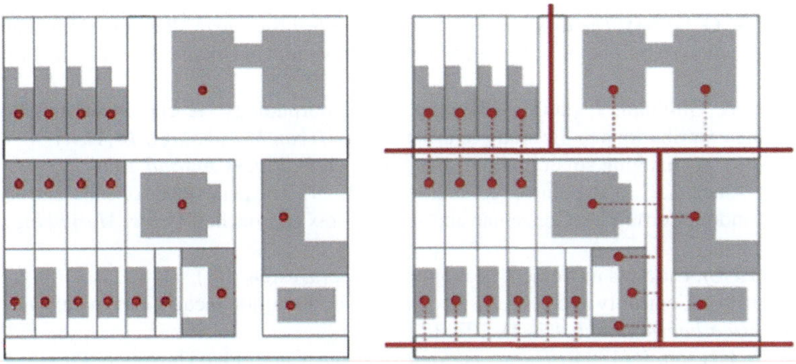

Fig. 3.8 Left: attractions like land use or number of shops are loaded as place-data on either plots or address points. Right: accessibility within the street network to different forms of attractions

Moreover, the interconnection of research and practice in urban morphology has strong roots. Both the logical and empirical bases for the link are evident (Whitehand, 2005, 2013). Yet the synergy between urban morphology and urban planning and design is still lacking (Barke, 2018). The study of urban landscape development does not reflect a lack of interest in history and the built environment among designers and planners, such as the designation of conservation areas and listed buildings. More significant is the lack of awareness of the wider historico-geographical structure of urban landscape.

Urban morphology could strengthen the historical awareness of urban designers and planners and provide them with evidence-based approach to manage change in historic city and areas. That indicates the necessity of an integrated methodological framework that provides operational support with a specific plan or strategy to practitioners and help them achieving a balance between maintaining the identity of local areas and city as a whole and producing forms of development that are attractive and marketable. Therefore, testing and applying the integrated approach in different urban contexts could be a focus for future studies.

References

Aymonino, C., Brusatin, G., Fabbri, M., Lena, P., Loverro, S., Lucianetti, & Rossi, A. (1966). *La Citta di Padova, saggio di analisi urbana*. Officina edizoni.

Barke, M. (2018). The importance of urban form as an object of study. In V. Oliveira (Ed.), *Teaching urban morphology* (pp. 11–30). Springer.

Bosselmann, P. (2008). *Urban transformation: Understanding city design and form*. Island Press.

Chen, F., & Thwaites, K. (2013). *Chinese urban design: The typomorphological approach*. Ashgate Pub. Ltd.

Conzen, M. R. G. (1960). Alnwick, Northumberland: A study in town-plan analysis. *Transactions and Papers (Institute of British Geographers)*, iii-122.

Conzen, M. R. G. (1964). *The urban morphology of Central Newcastle*. Paper presented at the the XXth International Geographical Congress 1964, Symposium in Urban Geography S2 Edinburgh, Scotland.
Gauthier, P., & Gilliland, J. (2006). Mapping urban morphology: A classification scheme for interpreting contributions to the study of urban form. *Urban Morphology, 10*(1), 41.
Gehl, J. (1989). A changing street life in a changing society. *Places Journal, 6*(1).
Griffiths, S., Jones, C. E., Vaughan, L., & Haklay, M. (2010). The persistence of suburban centres in Greater London: Combining Conzenian and space syntax approaches. *Urban Morphology, 14*(2), 85–99.
Hillier, B. (1996). Cities as movement economies. *Urban Design, 1*(1), 41–41.
Hillier, B. (1999). Centrality as a process: Accounting for attraction inequalities in deformed grids. *Urban Design International, 4*(3–4), 107–127.
Hillier, B., & Hanson, J. (1984). *The social logic of space*. Cambridge University Press.
Jacobs, J. (1961). *The death and life of great American cities*. Wiley.
Karimi, K. (1999). Urban conservation and spatial transformations. *Paper presented at the The Space Syntax second international symposium proceedings*, Brasillia.
Karimi, K. (2012). A configurational approach to analytical urban design: 'Space syntax' methodology. *Urban Design International, 17*(4), 297–318.
Kropf, K. (1993). *An enquiry into the definition of built form in urban morphology*. (Ph.D). University of Birmingham, UK.
Kropf, K. (2009). Aspects of urban form. *Urban Morphology, 13*(2), 105–120.
Kropf, K. (2011). Morphological investigations: Cutting into the substance of urban form. *Built Environment, 37*(4), 393–408. https://doi.org/10.2148/benv.37.4.393
Kropf, K. (2014). Ambiguity in the definition of built form. *Urban Morphology, 18*(1), 41–57.
Kropf, K. (2017). Bridging configurational and urban tissue analysis. *Paper presented at the Proceedings, 11th International Space Syntax Symposium*, Lisbon.
Kropf, K. (2018). *The handbook of urban morphology*. Wiley.
Larkham, P. J. (2006). The study of urban form in Great Britain. *Urban Morphology, 10*(2), 117–141.
Levy, A. (2005). New orientations in urban morphology. *Urban Morphology, 9*, 50–53.
Li, Y., & Gauthier, P. (2014). The evolution of residential buildings and urban tissues in Guangzhou, China: Morphological and typological perspectives. *Urban Morphology, 18*(2).
Lynch, K. (1984). *Good city form*. MIT Press.
Maffei, G., & Whitehand, J. (2001). Diffusing Caniggian ideas. *Urban Morphology, 5*(1), 47–48.
Monteiro, C., & Pinho, P. (2021). MAP: A methodology for morphological analysis and prescription. *Urban Morphology, 25*(1), 57–75.
Montgomery, J. (1998). Making a city: Urbanity, vitality and urban design. *Journal of Urban Design, 3*(1), 93–116.
Moudon, A. V. (1994). Getting to know the built landscape: Typomorphology. In *Ordering space: Types in architecture and design* (pp. 289–311).
Moudon, A. V. (1997). Urban morphology as an emerging interdisciplinary field. *Urban Morphology, 1*(1), 3–10.
Oliveira, V. M., & Claudia, P. J. (2015). A comparative study of urban form. *Urban Morphology, 19*(1), 73–92.
Pinho, P., & Oliveira, V. (2009). Different approaches in the study of urban form. *Journal of Urbanism: International Research on Placemaking and Urban Sustainability, 2*(2), 103–125. https://doi.org/10.1080/17549170903083676
Stahle, A., Marcus, L., & Karlström, A. (2005). Place syntax: Geographic accessibility with axial lines in GIS. *Paper presented at the fifth international space syntax symposium*, Delft.
Whitehand, J., & Gu, K. (2010). Conserving urban landscape heritage: A geographical approach. *Procedia-Social and Behavioral Sciences, 2*(5), 6948–6953.
Whitehand, J., Gu, K., Whitehand, S. M., & Zhang, J. (2011). Urban morphology and conservation in China. *Cities, 28*(2), 171–185.

Whitehand, J. W. (2019). ISUF and urban morphology: 25 years on and counting. *Urban Morphology, 23*(1), 103–104.
Whitehand, J. W. R. (2005). Urban morphology, urban landscape management and fringe belts. *Urban Design, 93*, 19–21.
Whitehand, J. W. R. (2012). Issues in urban morpholgy. *Urban Morphology, 16*(1).
Whitehand, J. W. R. (2013). Urban morphological research and practice. *Urban Morphology, 1*(17), 3–4.
Whitehand, J. W. R. (2015). Seeking an integrated approach to urban form: Tasks ahead. *Urban Morphology, 19*(1), 3–4.
Whitehand, J. W. R. (2017). Bridging the gaps: Urban morphology 20 years on. *Urban Morphology, 21*(1), 3–4.
Whitehand, J. W. R., & Gu, K. (2007). Urban conservation in China: Historical development, current practice and morphological approach. *Town Planning Review, 78*(5), 643–670.
Whitehand, J. W. R., Gu, K., & Conzen, M. P. (2014). The typological process and the morphological period: A cross-cultural assessment. *Environment and Planning b: Planning and Design, 41*(3), 512–533.
Zhang, Y., & Li, X. (2020). What new insights can the combination of the historico-geographical and configurational approaches to urban morphology offer? *24, 1*, 110–112.

Part II
Conceptualization and Evaluation

Chapter 4
An Integrated Approach: Combining the Historico-Geographical and the Configurational Approaches

Abstract This chapter goes one step further debating how the historico-geographical and configurational approaches can be combined to offer better description, explanation and prescription of the physical form of the built environment. A conceptual framework is developed to understand how and why urban form changes on the one hand, and how urban form conditions and affects urban lives on the other hand. In other words, this chapter brings an insight into the character of urban form in terms of its historicity and spatial potentials in generating socio-economic vibrancy, as the two sides of a coin. A methodological framework is developed to firstly explore how the integrated approach could contribute to plan unit recognition by systematically relating the results of both approaches, that is, the integration-segregation patterns of the street network with the distribution of hierarchical plan units. Secondly, it explores how the integrated approach could evaluate landscape unit based on the historical expressiveness and spatial potentials of urban form. It is anticipated that this evaluation could provide guidance for future management of historical urban landscape regarding to what to change and what to maintain and contribute to a synergy between conservation and redevelopment.

Keywords The historico-geographical approach · The configurational approach · Plan unit recognition · Landscape unit evaluation

4.1 Conceptual Framework

An important point about urban landscapes is that they change: 'the landscape we see is not a static arrangement of objects. It has become what it is, and it is usually in the process of becoming something different' (Darby, 1953). Underpinning the book is the view that cities as dynamic systems continue to grow and accommodate ever-changing human functioning while maintaining the historicity expressed by their distinctive physical fabric. It is the physical form of the built environment that constitutes the object of analysis in addressing the relationship between space and society. Thus, character of urban landscape in this book is about the historical interpretations of urban form. It is a dynamic concept—evolution and change are fundamental characteristics of cities. Character is also about people. It involves

X. Li and Y. Zhang, *Conserving and Managing Historical Urban Landscape*,
The Urban Book Series, https://doi.org/10.1007/978-981-19-4222-8_4

having an understanding of the activity, use, and movement within and through a place and its connections and linkages to its surroundings.

Another important point about the built environment is that the organization and transformations of urban form are not random but follow particular laws and logic (Levy, 1999). In other words, urban form is understood as a system of its own that is governed by the internal logic. At the same time, urban form is shaped by social, economic, cultural forces and is the result of deliberate and coordinated human effort (Kropf, 2009). In this book, the built environment is conceptualized as a system granted with relative autonomy, which is also the common object of enquiry of different morphological approaches (Gauthier & Gilliland, 2006).

Both of the historio-geographical and configurational approaches could provide explanations for how and why urban form changes. The historico-geographical approach is more from an external perspective, focusing on the physical development of the city conditioned by social, economic and cultural factors. The configurational approach is from an internal perspective, uncovering the generic logic for the process of a city's physical formation and transformation. Both approaches could prescribe what urban forms to be built. The historico-geographical approach could provide evidence for area delimitation and areas to change and maintain in old landscapes, while the configurational approach could propose spatial interventions to improve socio-economic performance (Fig. 4.1).

This book focuses on these two dominant approaches in urban morphology and goes one step further debating how these two approaches can be combined to enable a better understanding of the historical formation and transformation of urban form through plan unit delimitation. Plan unit is the most resistant to change among the three *form complexes* (ground plan, building fabric and land use) and encapsulates the most information of the historical stratification of a town. Moreover, it also

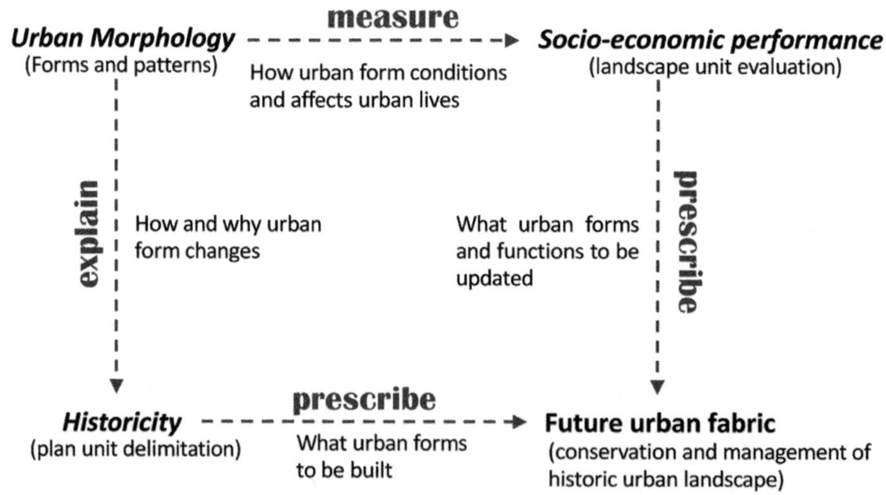

Fig. 4.1 Conceptual framework (*Source* Authors)

explores how these two approaches could be combined to evaluate the character and strength of urban landscape unit in terms of its historicity and potentials for generating socio-economic vibrancy, so that informed decisions can be made to deliver socially and economically sustainable outcomes. Landscape unit is delimited by overlaying the previous plan unit analysis with the land and building utilization. Though they tend to more ephemeral, the land and building utilization constitute an important aspect of societal activities that can be observed on the ground. The theoretical and methodological frameworks are elaborated in this section.

4.1.1 Historicity: Plan Unit Delimitation

As discussed in Sect. 3.1, further investigations are needed to elucidate to what extent and in which way the historico-geographical and configurational approach can be combined, and how the combination can contribute to enhancing the understanding of the historical transformation of urban form. The street is the primary focus of this study, as Oliveira et al. (2015) argue that the historico-geographical and configurational approaches are interrelated by the common denominator—street, which is the key built form element for delimiting plan units and recognizing accessibility. Building on existing endeavours, our study is an initial effort to inform the direction of this research. The research questions are: could street configuration contribute to plan unit recognition, and if so, to what extent? Could the historico-geographical process of a settlement explain the impact of street configuration on the transformation of plan units, and if so, to what extent?

Through systematically examining and comparing urban form across different morphological periods, on the one hand, knowledge about the changes and continuities of street configuration can be applied to explaining the growth of, and the interrelationship between plan units. This is premised on that the configurational approach, in particular 'space syntax', sees street configuration as encapsulating generative rules of space (Hillier, 1996, 2007), and to a large extent, accounts for the spatial distribution of pedestrian movements (Hillier et al., 1993) and subsequently different land uses (Hillier, 1996). Moreover, the configurational measures usually uncover some intrinsic spatial structures of urban areas that are more likely to be associated with historical urban processes, which may be related to the emergence of plan units. For instance, the integration (as in space syntax) patterns of street network have been identified and proved correlated with the spatial distribution of economic activities in different cities (e.g. Al Sayed et al., 2009; Porta et al., 2009, 2012; Yang & Hillier, 2007). It can be expected that the configurational properties of the street network could uncover the intrinsic spatial structures of urban areas that are likely to be associated with the historical urban process and potentially offer an additional reference for plan unit delimitation.

On the other hand, the place-specific knowledge about the historical process of the settlement can be employed to interpret the configurational measures (integration values) of plan units and examine the impact of street configuration on the emergence or transformation of plan units—for instance, to what extent the spatial-functional relationship dominates and to what extent it subordinates to other forces in the locality.

It can therefore be seen that combining the two approaches has the potential to provide an enhanced understanding of the physical transformation of urban form from both external and internal perspectives. That is physical form is conditioned by the external socio-economic factors, while at the same time, it is also underscored by the more generic generative rules of urban space. The extent to which these two aspects influence the transformation of urban form is dependent on the different development process of urban areas—incremental growth or fast-changing urban landscapes. In traditional historical towns, the position and alignment of buildings and plots are in close relation to the street, which forms the basis of traditional urban development. While in cities with fast-changing urban landscape, especially the new built-up areas, the relation between buildings and streets is relatively loose, and the modern architecture usually has a free-standing ground. The traditional typological configuration of street has been greatly eliminated. It is hypothesized that the street configuration plays a more effective role in plan unit recognition in traditional towns than in fast-changing urban landscapes. This hypothesis will be verified through empirical studies with contrasting urban contexts, and the robustness of this integrated morphological framework in elaborating the historical process of urban form will be explored in Chap. 5.

4.1.2 Historicity and Socio-economic Vibrancy: Landscape Unit Evaluation

As mentioned in Sect. 3.2, the built environment is not only a container of past history and development, it is also an inherent part of humans and their contemporary activities, where physical form is considered as a medium for a deeper understanding of how people live, work and play in cities. Jacob (1961)'s argument that a mixture of primary uses within each district and good spatial accessibility is prerequisites for creating 'vibrant' neighbourhoods against the separation of living, working and recreation in modernist city planning. In this research, the socio-economic profile of urban form is defined as the flow of people that all have different destination moving through the built environment, including pedestrian movement on street and people's activity occurring in buildings. Under this premise, the underlying urban image could be envisaged, and areas with mixed activities and good spatial accessibility are more likely to be socio-economically vibrant. Thus, the primary method of quantifying the socio-economic vibrancy of urban form in this book is the measurement of two spatial attributes: functional diversity and spatial accessibility.

Functional diversity refers to multiple uses (living, working and shopping) in the fine grain (Jacobs, 1987; Lynch, 1984; Bosselmann, 2008). It is defined in terms of function and form. The function mixture requires a wide diversity of ingredients, such as cafes, grocery stores, shops, cinemas, galleries and clubs within easy travelling distance. The diversity of business types is a preventive measure against economic downturns (Montgomery, 1998). Meanwhile, a mix in the district level may not mean there is a mix on the building level and vice versa. The scale level and the grain size of mixed activities vary from a mix on a building level to the level of the block. Thus, the scale level needs to be considered in mixed use. With the help of ArcGIS and emerging new datasets, a quantitative approach of measuring and describing mixed use at the scale of buildings emerged (Hoek, 2008, 2010). The assumption of this approach is that the mixed use organized on a walkable scale in cities or urban areas help creates socio-economic vibrancy. The mixed use is defined as the combination of three primary urban uses: live, work and visit, to which all urban functions could be related. In this way, the complexity of the functional mix in an urban district can be reduced to a simple index—mixed-use index (MXI) according to the functional balance between the proportions of living, working and visiting uses within a given urban district. The different proportions also indicate the liveliness of an area during the day and at night (Hoek, 2010). For example, the amount of amenities influences the number of visitors and flows in an area during both daytime and night time; the amount of working influences the number of workers in an area in the daytime; the amount of housing influences the number of residents in an area in the night time. If an area has equal amounts of each of these functions, a lively and diverse environment can be expected.

Spatial accessibility is seen as the generator of movement patterns, which in turn shape patterns of potential copresence and encounter. They provide the basic conditions for social events to be generated on street, and ultimately, vibrant street life (Hillier, 1989, p. 13). The configurational approach (space syntax) is concerned with the probabilistic effect on patterns of movement, which is based on statistical exploration between movement and syntactic properties of street network (Hillier, 1996; Hillier et al., 1993). Empirical studies have demonstrated that the street configuration is a major determinant of movement flows. The geometric and topological structures of street configuration are critically involved in people's cognition and navigation of streets (Hillier & Iida, 2005; Hillier et al., 1993).

Moreover, with the help of available open data in recent years, configurational analysis (place syntax) allows more accurate and intuitive description and analysis of access to existing socio-economic activities of each place, which is also known as 'accessible functional diversity' (Marcus, 2007). This measure could capture the 'mixedness' of socio-economic activities that depend on the scale of perceiving by adopting a certain radius (for example, walking distance) from a particular location through the street network. It focuses on how spatial form structures 'people and things' in space, rather than particular functions located in space. It is closer to a person's actual perception of urban functions while walking along the streets rather than the conventional area-based measurements (Marcus, 2007, 2010; Marcus et al., 2017).

Moreover, based on the relationship between street configuration and movement, empirical studies have shown that a place with high spatial accessibility tends to have high movement potentials for attracting economic activities (Hillier et al., 1993;

Hillier et al., 1996, 2007; Hoek, 2010; Griffith et al., 2010). Thus, spatial accessibility measured by the configurational approach could uncover the locational potential of a place for supporting or impeding socio-economic activities. Building on this form-function relationship, it is anticipated that combining the two attributes—spatial accessibility and accessible functional diversity—could provide an opportunity to discern urban areas with varied spatial potentials (Fig. 4.2). For example, an area with good functional mix but poor spatial accessibility, its spatial potential for generating socio-economic vibrancy is defined as low because the low movement flows may not support its good functional mix in the long term. While such a measure does imply that the higher accessibility and accessible functional diversity is, the higher spatial potential is, it does not imply that a higher spatial potential is always better. It indicates whether certain interventions would create greater potential for generating socio-economic vibrancy. To what extent the potential could be improved would depend on the specific context and circumstances as well as the particular design and planning objectives.

It is also worth noted that this study does not argue that spatial configuration is the determinant factor in land use distribution. Other factors, such as zoning and planning regulations, or socio-economic development strategies, may outweigh the impact of street network structure on land use. It is looking at the form-function relationship to explore the potentials of urban form to support or impede socio-economic activities. These potentials may correspond with or oppose social, economic or political factors. Understanding this will provide an understanding of the city's inner form logic.

As mentioned in Sect. 3.2, the historico-geographical approach includes patterns of land use and building utilization, yet the city is viewed as 'a long-term asset whose importance extends far beyond its contemporary functional value' (Whitehand, 2007). The extent to which urban form supports or impedes vibrant urban life is not its focus. The configurational approach (space syntax and place syntax) could complement the historico-geographical analysis by studying the relationship between street network and movement patterns, as well as the impact on land use distribution. It could measure the impact of not only street configuration but also attractions on people's actual use and experience of space. Therefore, combining

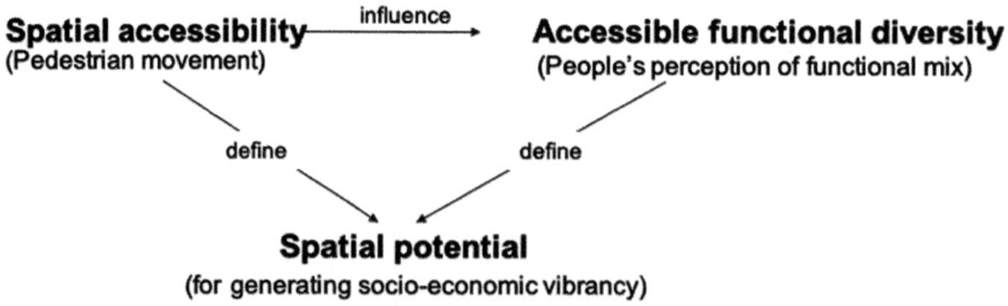

Fig. 4.2 Relationship between spatial accessibility, accessible functional diversity and spatial potential (*Source* Authors)

the two morphological analyzes could enable a fine-grained, cross-scale examination of the historical roots and socio-economic character of urban form, laying out and making more explicit both the opportunities and risks. For example, it may identify not only which areas shall be preserved because of their distinct historical expressiveness embodied in the physical environment, but also which areas tend to have comparative advantages in terms of socio-economic viability, and which areas are more likely to be confronted with decline that may lead to rapid deterioration of the physical environment and consequently loss of historical expressiveness. In addition, this combination can provide an understanding of how different morphological regions (landscape units) or plan units are spatially and socially connected and evaluate how changes to one may affect another across the urban grain.

This evaluation potentially allows for sharper identification of conservation priorities and insightful anticipation of inter-area influences, in order to achieve smart distribution and allocation of limited resources across urban grain. This integrated analysis provides a sound basis for suggesting ways of improving socio-economic performance of urban areas, such as improving spatial accessibility or injecting more diverse functions. It has the potential to make positive use of the patterns of formation when planning new environments and settlements, achieving a synergetic relationship between conservation and urban redevelopment (Fig. 4.3).

To sum up, this book started as historical research on urban transformation that was intended to understand urban form through its historical development. It looks both backward and forward at the same time, studying the past to understand the present with a strong perspective on the future. The integrated methodological framework of recognizing plan unit and evaluating landscape unit is introduced in the next section. How this analytical framework is applied in urban conservation, and management is discussed in detail in Part III.

Fig. 4.3 Complementary use of the historico-geographical and configurational approaches (*Source* Authors)

4.2 Methodological Framework

This section elaborates the methodological framework of combining the hisotrico-geographical and configurational approaches. Specific methods for analyzing the built form have been well developed, respectively, in both approaches. This methodological framework is proposed to more systematically relate the results of both approaches, i.e. the integration-segregation patterns of the street network with the distribution of hierarchical plan units. Nevertheless, it starts with separate historico-geographical and configurational analyzes of the built form of a settlement across its key morphological periods. As the present landscape is an accumulated record of distinct morphological periods, it provides a clearer understanding of the structure and dynamics of the built environment by successive cross-sections in time than only focusing on the current surviving urban landscape.

4.2.1 Plan Unit Recognition

4.2.1.1 Plan Unit

In this integrated framework, the historico-geographical analysis focuses on plan unit recognition through cartographic analysis (Gu & Zhang, 2014; Pinho & Oliveira, 2009; Whitehand, 2007). As ground plan is the most resistant to change among the three *form complexes* (ground plan, building fabric and land use), it encapsulates the most information of the historical development and transformation of a town (Conzen, 1960). Due to the same reason, urban form data—street, plot and building block—for developing accurate ground plans of a settlement across its morphological periods are most likely to be available. Particular emphasis is given to the building block-plan, instead of building types, building materials and architectural style, as the broader concern is the spatial structure change of the urban ground plan.

Tactically, cartographic data of the settlement such as historical and cadastral maps showing individual property boundaries, streets and building blocks of each morphological period are firstly collected, reproduced and compiled in chronological order. These chosen maps are imported and georeferenced in ArcGIS. With the aid of transparent overlay in ArcGIS, the extension, replacement or demolition of form elements could be rigorously identified. The historical maps, in conjunction with a plot-by-plot survey of the urban landscape, provide a sound basis for plan unit recognition for each morphological period.

Hierarchies of plan units are drawn as polygons or polylines in ArcGIS. The first-order boundaries often distinguish areas with relative homogeneity in the same morphological period, such as the old town, or the residential accretions. The second-order boundaries identify the subdivisions within the town. The third-order boundaries often distinguish the plan type, such as ribbon development along a major road. The recognition of plan unit provides a finer-scale division of urban areas, suggesting

Table 4.1 Configurational measures for analyzing the global and local structures of street network (Hillier et al., 2007, 2012)

Measures for the global structure	Measures for the local structure
NACH = log (Ch + 1)/log (TD + 3), where Ch = choice value TD = Total angular depth	Integration = 1/metric mean depth (MMD) MMD = Total metric length/(NC − 1), where NC = Number of segments

a detailed construction of the long-term historic development and transformation of historic cities or areas.

4.2.1.2 Spatial Accessibility

The configurational analysis explores the integration-segregation patterns of the street network at both the global and local scales. First, based on the above-mentioned ground plans, the street network models of the settlement under investigation are built up for each morphological period in ArcGIS. The series of models are then analyzed using configurational measures, as shown in Table 4.1. For the global scale, the measure employed is the normalized angular choice (NACH), which has been widely proved powerful in capturing the overall integration-segregation of the street network, and in facilitating comparisons between street networks of different sizes (Hillier et al., 2012). It is hypothesized that the overall integration-segregation pattern of the street network may help to explain the spatial distribution of plan units of the higher tier.

For the local scale, the inverse of mean metric depth (MMD) is chosen to capture the clusters of street segments that share similar configurational characteristics. As these clusters are usually correlated with distinctive urban activities and their associated built form features (building block and plot) (Hillier et al., 2007), it is hypothesized that the local patchwork patterns[1] of the street network could contribute to recognizing plan units of the lower tier. The local analysis is performed at the radii of 400 m and 800 m. As empirically this measure is highly sensitive to the size of the street network and dependent on the average street segment length, analysis conducted at both radii would allow for more accurate measurement and enhanced analysis.

4.2.1.3 Integrated Approach in Plan Unit Recognition

The results of the historico-geographical and the configurational analysis for each morphological period are then correlated (Fig. 4.4). The maps showing the global integration-segregation pattern and local patchwork pattern of the street network are

[1] Patchwork refers to metrically similar areas with homogenous identity in terms of the scale of grid structure (Hiller et al., 2007).

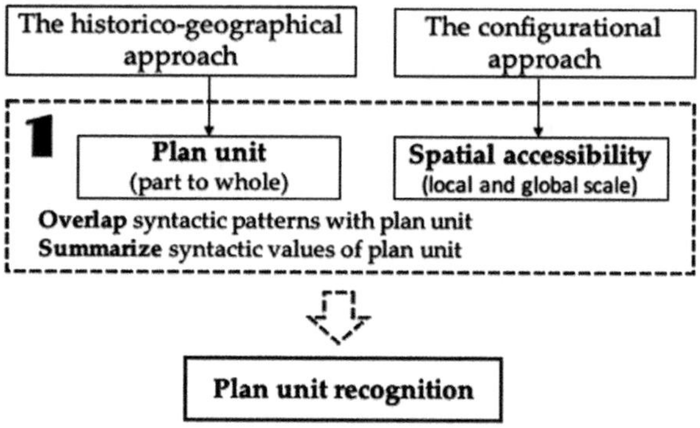

Fig. 4.4 Methodological framework for plan unit recognition (*Source* Authors)

overlaid to those of plan units. Their relationships are then investigated based on both the more intuitive maps and the more accurate descriptive statistics.

In the overlapped maps, the specific focus of the integrated analysis is on how the overall integration-segregation pattern relates to the layout of the higher-tier plan units, and how the local patchwork pattern corresponds to the distribution of the lower-tier plan units. It is anticipated that the results could help explain the extent to which the configurational approach could provide additional evidence for plan unit recognition besides conventional mapping of the street, plot and building block.

In the descriptive statistics, the aggregate values of both the global and local configurational measures are calculated for plan units in different hierarchies and compared across different morphological periods. The different configurational values of plan units uncover their varied spatial locations within the city and the potential spatial impact on the distribution of socio-economic activities in the historical context. With the help of the historico-geographical approach, the context-specific information about the historical process of the settlement is employed to examine the extent to which these spatial analyzes are socio-spatially meaningful.

In sum, combining the two approaches aims to provide enhanced understandings of the historical transformations of urban form. That is, the formation and transformation of plan units are not only conditioned by different social, economic and political factors but also affected by the generic generative rules of urban space.

4.2.2 Landscape Unit Evaluation

This methodological framework is proposed to more systematically relate the results of both approaches, i.e. the spatial potentials of urban form measured by the configurational approach with the distribution of hierarchical landscape units.

4.2.2.1 Landscape Unit

Apart from the plan unit division, the historico-geographical analysis also focuses on landscape unit identification. As discussed in Sect. 2.1.1, integrating the analysis of three form elements yields hierarchies of landscape units (Conzen, 1988). The chosen maps in each morphological period are georeferenced to the accurate plan of the most current situation in ArcGIS. Building types in terms of building age or function could be inferred from recent survey maps. The available land use data were collected from the planning authority. Plot as the unit of land use could help identify patterns of land uses, such as commercial or residential uses. These maps are compared with the aid of transparent overlays in ArcGIS to identify different hierarchies of landscape units. Based on the previous plan unit analysis, building fabric shows a moderate resistance to change, it determines mainly the intermediate rank within the plan units; land and building utilization tend to be more ephemeral, and they contribute to the low rank. The landscape unit delimitation enables a systematic understanding of the different degrees and types of historical expressiveness embodied in the urban landscape. Various parts of cities or districts are distinguished and characterized at different scales of analysis, from the individual building to entire settlement.

4.2.2.2 Configurational Analysis of Socio-economic Vibrancy

Accessible functional diversity

The configurational analysis focuses on measuring the two spatial attributes—functional diversity and spatial accessibility—to quantify the socio-economic vibrancy of urban form. To measure functional diversity, Hoek (2010) and Dovey and Pafka (2017)'s functional mix frameworks are adopted. Three primary urban functions are classified as live, work and visit. Live function includes residence and hotel; work function has university, school, office, industry and factory; visit function includes church, theatre, library, museum, restaurant and shop. The underlying assumption of this approach is that amount of specific activities in an area generally corresponds with the amount of specific floor areas. Thus, building floor areas are taken as a proxy for the amount of socio-economic activities. The different types of functional mix are measured by the different proportions of floor areas of live/work/visit. As shown in Fig. 4.5, a simplified colour coding of functional mix framework is adopted to represent three primary functions (live/work/visit) and various types of functional mix—mono-function, bi-function and mixed-function—in the triangle diagram.

Emerging point of interest data (POIs), which carry the spatial locations of individual attractions at building scale, is collected from online maps and imported in ArcGIS. It could contribute to land use data by uncovering the functional mix within buildings, either vertically or horizontally based on the above-mentioned survey maps. According to this framework, POIs and land use are reclassified into live/work/visit categories (Tables 4.2 and 4.3). Floor areas of live/work/visit functions are measured and distributed to physical plot, which is the common form element and

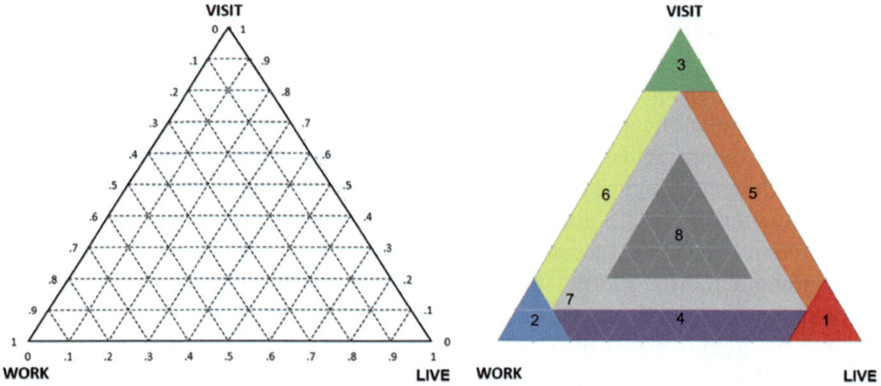

Fig. 4.5 Tertiary diagram adopted in this study. Left: Tertiary diagram showing the coordinates based on the proportion of floor areas of live/work/visit, ranging from 0 to 1. Right: Triangular diagram showing the different types of functional mix. Red–mono-live (>80%); blue–mono-work (>80%); green–mono-visit (>80%); purple–live/work (visit <10% while 10% < live, work <80%); orange—live/visit (work <10% while 10% < live, visit < 80%); yellow—visit/work (live < 10% while 10% < visit, work < 0%); grey—live/work/visit (each over 10%); dark grey—live/work/visit (each over 20%) (Hoek, 2010)

analytical unit in both the historico-geographical and configurational approach. This increases the analytical resolution and provides a finer-scale description and analysis of socio-economic activities in space. Land use of parks, green spaces, squares and vacant lands is not included as they could not be quantified by building floor areas. The detailed calculations will be elaborated in the specific case study.

As discussed previously in Sect. 2.1.3, conventional measure of functional mix is area-based, focusing on the descriptions and analysis of a single urban element — physical plot. However, functions are not limited by plot boundaries; instead, they could be reached by residents within a radius from a particular location. The accessible functional diversity allows more accurate and intuitive descriptions and analysis of access to existing functional diversity of each place, which is close to a person's actual perception of urban fabric while walking along the streets. It could be measured by "attraction reach" in place syntax (Stahle et al., 2005, 2012). The sum of the reachable functions is proportional to the attractiveness of destinations and inversely proportional to the distance in between. Attraction reach (AR) for a given origin i is calculated as.

$$AR(i) = \sum_{i \in I} (f(j)w(D(i, j)))$$

Where

I the rest of reachable attractions within given radius
$f(j)$ attraction value associated with attraction j
$D(i,j)$ shortest distance from origin i to attraction j.

Table 4.2 Reclassification of land use under Hoek (2010) and Dovey and Pafka (2017)'s framework

Land use division	Subdivision			New category
(R) Residence	R1, R2, R3		Residence	Live
	Ra	Raa	Student hostels	
		Rab	Elderly hostels	
	Rb		Mixed residence with commerce	
(B) Commerce and service	B14		Hotel	
(A) Public facility	A1		Government institution	Work
	A3	A31	University and military college	
		A32	Polytechnic school	
		A33a	Primary school	
		A33b	Secondary school	
		A33c	High school	
		A34	Special education school	
		A35	Research institute	
(B) Commerce and service	Bb		Mixed office with commerce	
(M) Industry	M1, M2, M3		Industrial sites	
(R) Residence	Rc		Community centre	Visit
(A) Public facility	A2	A21	Library, museum, gallery, exhibition hall	
		A22	Youth and elderly activity centre	
	A4		Indoor gymnasium	
	A5		Hospital	
	A7		Heritage site	
	A9		Religious site	
(B) Commerce and service	B1	B11	Retail, mall, market	
		B13	Catering	
	B2	B21	Insurance, bank, stock	
		B22	Television production, media	
	B29a		Research institute	
	B29b		Consultant	
	B3	B31	Theatre, music hall, dance hall	
	B4		Postal office, telecom service	
	B9		Vehicle repair shop	

Table 4.3 Reclassification of POIs under Hoek (2010) and Dovey and Pafka (2017)'s framework

POI category	Subcategory	New category
Residence	Residential units	Live
Hotel	Hotels	
Government agency	Government agency	Work
School and institution	School	
	Education organization	
	Research institute	
Business office	Business office	
Cultural facilities	Museum, library	Visit
Leisure and sports	Sports hall	
	Ktv, Majiang, etc.	
Medical service	Pharmacy	
	Hospital	
Catering	Restaurant	
	Coffee shops, teahouse	
Financial services	ATM	
	Bank	
Amenities	Law firms	
	Bath centre	
	Hair and beauty salon	
	Customer service centre	
	Laundry	
	Travel agency	
	Real estate agency	
Shopping	Vegetable market	
	Supermarket	
	Retails	
	Stores of household appliances	
	Shopping mall	
Service enterprise	Service enterprise	

Table 4.4 Distance decay function of 'attraction reach' in place syntax (Stahle, 2012)

Measurement	Line model	Distance	Radius	Distance-decay function
Attraction reach	Dual model	Metric	400 m	$Y = 1 - x^{\wedge}1$ (linear function)

$W(x)$ attenuation function, the exponent for adjusting the effect of distance decay in the calculation. The corresponding shape of distance decay can be derived from the assumed model of travel.

The given radius in this study is limited to a walkable scale of 400 m (a threshold of 5-min walking distance). It is the most effective radius setting according to extensive research in urban studies (Hillier, 2009). The distance decay describes the effect of distance on functional interactions between buildings. It states that the functional interactions between two buildings decline as the distance between them increases. The distance decay function and exponents for adjusting the decay effect are put in Table 4.4. X refers to the distance, and Y refers to the weight of attract. Metric distance in PST is a linear function.

4.2.2.3 Spatial Accessibility

Spatial accessibility as the other spatial attributes is measured by space syntax to predict urban movement on different scales (Table 4.5). In the global scale, the normalized least angle integration (NAIN) is chosen as the syntactic measure, as it is powerful in both analysis and movement prediction and is capable of providing a deeper understanding of the spatial morphology of cities compared to the previous least angle integration measures (Hillier et al., 2012). In the local scale, the inverse of mean metric depth (MMD) is chosen because it is able to capture locally walkable areas (Hillier, 2007).

Empirical studies have found that more integrated locations are more attractive as destinations for movement-reliant socio-economic activities based on the relationship between street network and 'to movement' (Hillier, 1993). The extent to which spatial accessibility is related to functional mix is explored at the scale of plot. It is hypothesized that the work and visit functions are more likely to be distributed along the globally accessible streets, while the residential function is more likely to be located along the locally accessible streets. As a result, the globally accessible physical plots tend to have high concentrations of work/visit functions and higher

Table 4.5 Spatial accessibility measures at the global and local scales (*Source* Authors)

Measures for the global accessibility	Measures for the local accessibility
NAIN = (NC + 2) ^1.2 /TD NC = total node count TD = Total depth	Integration = 1/metric mean depth (MMD) MMD = Total metric length / (NC−1) where NC = Number of segments

Table 4.6 Definition of high, medium and low values in spatial accessibility and accessible functional diversity using natural break method in ArcGIS (*Source* Authors)

Spatial accessibility	Classification
High	High values in global analysis
Medium	Medium values in global analysis
Low	Low values in global analysis
Accessible functional diversity	*Classification*
High	Mixture of three functions
Medium	Bi-functional
Low	Mono-functional

degrees of mix, while the locally accessible physical plots tend to have higher proportions of live function and lower degrees of mix. Thus, the global accessibility could reveal the locational potentials for diverse socio-economic activities. It is classified into high, medium and low based on the global NAIN values.

The accessible functional diversity of each physical plot is classified into three levels accordingly: high value represents mixed-function; medium value as bi-function and low value as mono-function. This high-to-low level underlies the ability of urban form to accommodating diverse activities (Table 4.6).

Spatial potential for generating socio-economic vibrancy

Overlaying these two parameters creates an opportunity to classify various types of urban areas according to their spatial potentials for generating socio-economic vibrancy, namely 'matched', 'in-between' and 'mismatched' (Table 4.7). 'Matched' typology reflects similar values in accessible functional mix and spatial accessibility (first row). This suggests the locational advantage of these areas has been fully made use of in supporting their current functional diversity. 'In-between' typology reflects the existence of a slight difference between these two variables (second row): either higher spatial accessibility, lower functional mix or higher functional mix and lower spatial accessibility. For areas of the former type, the relatively lower functional mix has not fully made use of their higher spatial advantage; for areas of the latter, the relatively lower spatial advantage may not support their higher functional mix in the long run. 'Mismatched' typology reflects the existence of significant differences between these two, either mono-use and high spatial accessibility or mixed use and low accessibility (third row).

This classification framework provide a systematic description and analysis of the variations of spatial potential of urban form. Clusters of physical plots with the same typologies of spatial potentials could be highlighted, suggesting a finer-scale subdivision of urban areas. Those 'in-between' and 'mismatched' urban areas contain at least one spatial metric with relatively low value, which also indicates their potentials to change in the future.

Table 4.7 Classification of spatial potentials based on the accessible functional diversity and the spatial accessibility (*Source* Authors)

Spatial accessibility	Accessible functional diversity	Spatial potential
High	High (mixed)	**Matched** high
Medium	Medium (bi-function)	**Matched** medium
Low	Low (mono)	**Matched** low
Medium	Low (mono)	**In-between** (higher spatial accessibility/lower functional mix)
High	Medium (bi-function)	
Low	Medium (bi-function)	**In-between** (higher functional mix/lower spatial accessibility)
Medium	High (mixed)	
High	Low (mono)	**Mismatched** (high spatial accessibility/low functional mix)
Low	High (mixed)	**Mismatched** (high functional mix/low spatial accessibility)

4.2.2.4 Integrated Approach in Landscape Unit Evaluation

The mappings of the historico-geographical and the configurational analysis are then overlaid to examine the spatial potentials of different hierarchies of landscape units (Fig. 4.6). Evaluations are based on the overlaid mappings, together with a detailed written description and photographs.

For each hierarchy of landscape units, its historicity and spatial potentials embedded in urban form are summarized and presented in Table 4.8. In the first column, a detailed description of the street, building block and plot patterns is provided, which helps to clarify what is important to the historic character and whether existing urban forms accord with the character of the urban landscape units to which they belong. In the second column, the spatial potential of landscape unit is presented in terms of spatial accessibility and accessible functional diversity. Spatial metrics with low values could be identified, for example, poor spatial locations which may not support the diverse content or areas with mono-function which may not fully take advantage of its spatial accessibility. In the third column, evaluations of the strengths and weaknesses of landscape units are carried out. For example, which parts of landscape units exhibit both distinguished historic character and good spatial potential that have comparative advantage for conservation, and which parts confront deteriorated historical expressiveness and poor spatial potential that may need physical improvement for future redevelopment.

This evaluation sets up a basis for future interventions in urban planning and design, providing reference for what to change and what to remain at different scales, from the individual plot to entire settlement. For the entire urban grain, it potentially allows insightful anticipation of inter-area influences, in order to achieve smart distribution and allocation of limited resources. For a specific urban area, it subdivides urban areas into smaller clusters, allowing for sharper identification of conservation and redevelopment priorities.

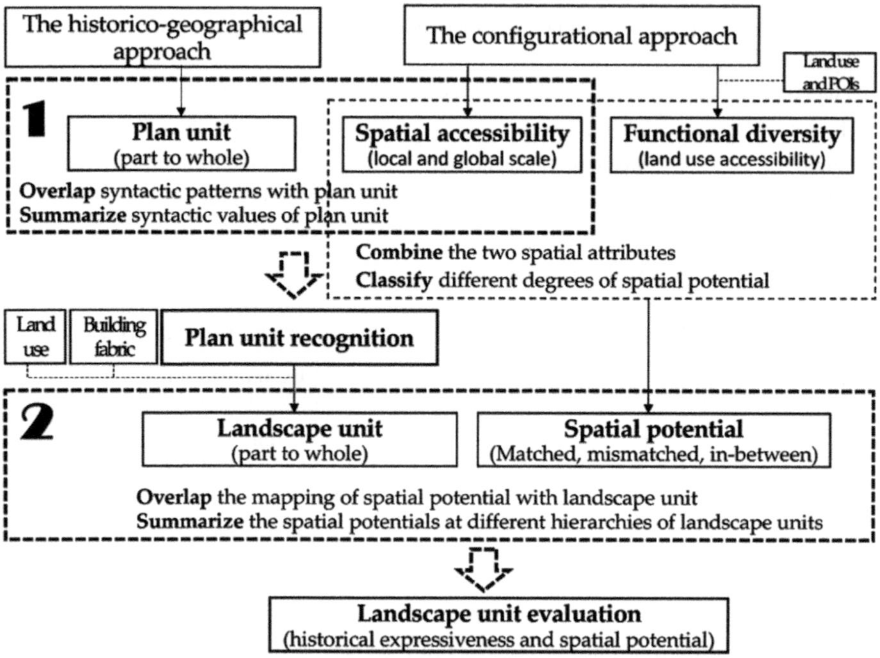

Fig. 4.6 Methodological framework for landscape unit evaluation (*Source* Authors)

To sum up, the conceptual and methodological frameworks are presented in this chapter to explain the historical transformations of urban form and to evaluate its the strengths and weaknesses based on the historicity and spatial potentials for generating socio-economic vibrancy. They can be seen as an incipient step to addressing the challenge of coordinating and combining different morphological schools (Whitehand, 2012, 2015, 2017).

As discussed in Sect. 4.1, the extent to which the integrated approach could provide enhanced understandings of the historical transformation of urban form will be tested under different urban contexts. Ludlow, UK and Chinatown, Singapore are firstly analyzed and compared in Chap. 5. As Ludlow, UK (Conzen, 1988) is the object of one seminal work of the historico-geographical approach, and the concrete historico-geographical investigation on this case study offers a very sound basis for exploring the complementarities between these two morphological approaches. While Chinatown, Singapore is in striking contrast with the well-preserved historic town of Ludlow, it has a relatively shorter history starting from 1819 (Liu, 1995) and has undergone large-scale demolition and redevelopment in the past thirty years. Its urban planning has been characterized by the top-down approach. Chinatown is introduced following our hypothesis that if the first case study yields positive results, then this methodological framework that combines the two morphological approaches is likely to be effective to unpack the historical transformation of urban form in drastically different urban contexts. Nanjing, China is also selected to test this framework in

Table 4.8 Evaluation framework of landscape units (*Source* Authors)

Historicity						Spatial potential		Evaluation
Hierarchical landscape unit			Street	Plot	Building block	Spatial accessibility	Accessible functional diversity	Strengths and weaknesses
1st	I		Urban settlement of a certain period					
	2nd	Ia	Subdivisions within the settlement					
		3rd	Ia1	Plan types		High	Low	This area is characterized with significant historical features. Yet its spatial potential awaits further improvement. The mono-function may not fully take advantage of its spatial accessibility. The good spatial advantage has not been fully used.
		3rd	Ia2			Low	High	This area is characterized with significant historical features. Its poor spatial locations may not support its diverse content, which may affect the viability of socio-economic activities in the long run.

the first part of Chap. 6. Similar with Singapore, China has acute conservation problems associated with explosive economic growth and pressures for redeveloping its historical urban areas. The historic core of Nanjing has inevitably undergone great transformations since the economic reform in 1978. As these three cities were developed under different planning decisions and characterized with either incremental growth or fast-changing urban landscape, it is interesting to explore whether their urban layouts exhibit similar development process and follow the generic generative rules of urban formation and transformation.

To move on further, in the second part of Chap. 6, Nanjing is also selected out of the above three cases to further apply the combined morphological analysis. Like other typical historic cities in China, the unprecedented pace in the change of urban landscape has presented Nanjing with new challenges for the comprehension and management of urban forms in the urban revitalization and redevelopment process. This integrated approach will assess the historical expressiveness and spatial potentials of urban form for generating socio-economic vibrancy. Consistently, building on the morphological analysis of Nanjing, Chaps. 7 and 8 in Part III will demonstrate how the analysis can inform conservation practice by zooming in on two historical urban areas in Nanjing, with a view to benefit wider application to other cities facing the conservation and regeneration dilemma.

References

Al Sayed, K., Turner, A., & Hanna, S. (2009). *Cities as emergent models: The morphological logic of Manhattan and Barcelona*. Paper presented at the 7th International Space Syntax Symposium, Stockholm.

Bosselmann, P. (2008). *Urban transformation: Understanding city design and form*. Island Press.

Conzen, M. R. G. (1960). Alnwick, Northumberland: A study in town-plan analysis. *Transactions and Papers (Institute of British Geographers)*, iii–122.

Conzen, M. R. G. (1988). Morphogenesis, morphological regions and secular human agency in the historic townscape, as exemplified by Ludlow. *Urban historical geography: Recent progress in Britain and Germany*, pp. 253–272.

Darby, H. C. (1953). On the relations of geography and history. *Transactions and Papers (Institute of British Geographers)* (19), 1–11.

Dovey, K., & Pafka, E. (2017). What is functional mix? An assemblage approach. *Planning Theory & Practice*, 1–19.

Gauthier, P., & Gilliland, J. (2006). Mapping Urban morphology: A classification scheme for interpreting contributions to the study of Urban form. *Urban Morphology, 10*(1), 41.

Griffiths, S., Jones, C. E., Vaughan, L., & Haklay, M. (2010). The persistence of suburban centres in Greater London: Combining Conzenian and space syntax approaches. *Urban Morphology, 14*(2), 85–99.

Gu, K., & Zhang, J. (2014). Cartographical sources for Urban morphological research in China. *Urban Morphology, 18*(1), 5–21.

Hillier, W. R. T., Yang, T., & Turner, A. (2012). Normalising least angle choice in Depthmap-and how it opens up new perspectives on the global and local analysis of city space. *Journal of Space syntax, 3*(2), 155-193.

Hillier, B. (1989). The architecture of the Urban object. *Ekistics, 56*, 5–21.

Hillier, B. (1996). Cities as movement economies. *Urban Design, 1*(1), 41–41.

Hillier, B. (2009). *Spatial sustainability in cities: Organic patterns and sustainable forms*. Paper presented at the 7th International Space Syntax Symposium, Stockholm.

Hillier, B., & Iida, S. (2005). Network and psychological effects in Urban movement. In A. Cohn & D. Mark (Eds.), *Spatial information theory* (Vol. 3693, pp. 475–490). Springer.

Hillier, B., Penn, A., Hanson, J., Grajewski, T., & Xu, J. (1993). Natural movement: Or, configuration and attraction in Urban pedestrian movement. *Environment and Planning B: Planning and Design, 20*(1), 29–66.

Hillier, B., Turner, A., Yang, T., & Park, H. T. (2007). *Metric and topo-geometric properties of Urban street networks*. Paper presented at the Proceedings Space Syntax. 6th International Space Syntax Symposium, Istanbul.

Hoek, J. V. D. (2008). The MXI (Mixed-use Index) as tool for Urban planning and analysis. *Corporations and Cities: Envisioning Corporate Real Estate in the Urban Future, Brussels*.

Hoek, J. V. D. (2010). The mixed use index (MXI) as planning tool for (New) Towns in the 21st Century. In M. Provoost (Ed.), *New towns for the 21st century: The planned versus the Unplanned City*. Martien de Vletter, SUN.

Jacobs, A., & Appleyard, D. (1987). Toward an Urban design manifesto. *Journal of the American Planning Association, 53*(1), 112–120.

Jacobs, J. (1961). *The death and life of great American Cities*. Wiley.

Kropf, K. (2009). Aspects of Urban form. *Urban Morphology, 13*(2), 105–120.

Levy, A. (1999). Urban morphology and the problem of the modern urban fabric: Some questions for research. *Urban morphology, 3*(2), 79–85.
Liu, G. (1995). *Pastel portraits: Singapore's Architectural Heritage.* Singapore Coordinating Committee.
Lynch, K. (1984). *Good city form.* MIT Press.
Marcus, L. (2007). *Spatial capital and how to measure it: An outline of an analytical theory of the social performativity of Urban form.* Paper presented at the Proceedings, 6th International Space Syntax Symposium, Istanbul.
Marcus, L. (2010). Spatial capital: A proposal for an extension of space syntax into a more general Urban morphology. *The Journal of Space Syntax, 1*, 30–40.
Marcus, L., Pont, M. B., & Bobkova, E. (2017). *Cities as accessible densities and diversities.* Paper presented at the 11th Space Syntax Symposium.
Montgomery, J. (1998). Making a city: Urbanity, vitality and Urban design. *Journal of Urban Design, 3*(1), 93–116.
Oliveira, V. M., Claudia; Partanen, Jenni (2015). A comparative study of Urban form. *Urban Morphology, 19*(1), 73–92.
Pinho, P., & Oliveira, V. (2009). Different approaches in the study of Urban form. *Journal of Urbanism: International Research on Placemaking and Urban Sustainability, 2*(2), 103–125. https://doi.org/10.1080/17549170903083676
Porta, S., Latora, V., Wang, F., Rueda, S., Strano, E., Scellato, S., Cardillo, A., Belli, E., & Cormenzana, B. (2012). Street centrality and the location of economic activities in Barcelona. *Urban Studies, 49*(7), 1471–1488.
Porta, S., Strano, E., Iacoviello, V., Messora, R., Latora, V., Cardillo, A., Wang, F., Scellato, S. (2009). Street centrality and densities of retail and services in Bologna, Italy. *Environment and Planning B: Planning and Design, 36*(3), 450–465.
Stahle, A. (2012). *Place syntax tool (PST).* Retrieved from https://www.accessibilityplanning.eu/uploads/pdf/19-PST-R.pdf
Stahle, A., Marcus, L., & Karlström, A. (2005). *Place syntax: Geographic accessibility with axial lines in GIS.* Paper presented at the Fifth international space syntax symposium, Delft.
Whitehand, J. W. R. (2007). *Conzenian Urban morphology and Urban landscapes.* Paper presented at the Proceedings, 6th International Space Syntax Symposium, Istanbul.
Whitehand, J. W. R. (2012). Issues in Urban morpholgy. *Urban Morphology, 16*(1).
Whitehand, J. W. R. (2015). Seeking an integrated approach to Urban form: Tasks ahead. *Urban Morphology, 19*(1), 3–4.
Whitehand, J. W. R. (2017). Bridging the gaps: Urban morphology 20 years on. *Urban Morphology, 21*(1), 3–4.
Yang, T., & Hillier, B. (2007). *The fuzzy boundary: The spatial definition of Urban areas.* Paper presented at the 6th International Space Syntax Symposium, Istanbul.

Chapter 5
Integrated Examination of Urban Form: Historicity

Abstract This chapter presents two case studies with contrasting urban contexts chosen to apply the integrated methodological framework in the study of the historical transformation of cities. They are Ludlow, a well-preserved historic town with incremental growth outward from the existing peripheries of the town in the UK, and Chinatown, a historic district that has undergone great urban redevelopment in the past decades in Singapore. The results show that on the one hand, the street configuration, especially the integration-segregation pattern, is correlated with the distribution of plan units and therefore can inform their recognition; on the other hand, the place-specific knowledge about the historico-geographical process of the city can help explain the changes in the configurational values of different plan units. The findings indicate that the combination of these two approaches is capable of providing enhanced understandings of the historical transformation of urban form for both urban areas with incremental growth and those with fast-changing urban landscapes.

Keyword Historico-geographical approach · Plan unit · Configurational approach · Space syntax · Street configurational · Ludlow (UK) · Chinatown (Singapore)

In this chapter, Ludlow, UK and Chinatown, Singapore are selected to test this methodological framework in sequence. The details of the case studies and analytical results are discussed separately. This piece of work was firstly published in Urban Morphology 25(1).

5.1 The Case Study: Ludlow, UK

Ludlow is a typical British market town in the UK, bordered by the River Teme to the west and south. Three morphological periods are identified: the mediaeval and early modern period (1270–1830), the Victorian and Edwardian period (1831–1918) and the modern period (1919–2015).

X. Li and Y. Zhang, *Conserving and Managing Historical Urban Landscape*, The Urban Book Series, https://doi.org/10.1007/978-981-19-4222-8_5

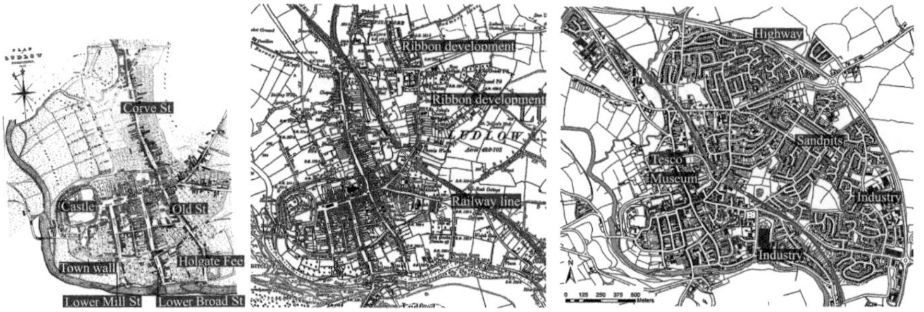

Fig. 5.1 Left. Map of Ludlow c. 1835 (*Source* Shropshire Archives). Middle. Map of Ludlow c. 1950s (*Source* old-maps.co.uk). Right. Map of Ludlow, 2015 (*Source* Ordnance Survey Website)

Across the three morphological periods, the castle dating back to between 1086 and 1094 remains almost unchanged. Within the thirteenth-century town wall, there have never been radical changes (Fig. 5.1 left). The planned grid structure has been well preserved, except for some amalgamations of burgages and changes in building types (Conzen, 1988). In the Victorian and Edwardian period (1830–1918), the population boom gave rise to the construction of new housing to the north and east of the mediaeval town, such as the terraced housing. They were constituted of standard plots and shallow rectangular building blocks along the major roads. This ribbon development represented Ludlow's first residential accretions in the Victorian and Edwardian period (Fig. 5.1 middle). In the modern period (1919–2015), periphery accretions took place much more extensively in the form of middle and working-class housing (Fig. 5.1 right). The new residential development largely filled the vacant land between the mediaeval town and the 1970s by-pass. Terraced housing became obsolete following World War I, and semi-detached housing appeared after the new legislation for suburban housing in 1919 (Scott, 2013; Whitehand, 2001). The large-scale industrial parks were constructed at the fringe of the highway and the Teme Riverside. Some large-scale modern building blocks, such as Tesco supermarket and Museum Resource Centre, were built at the northern fringe of the town wall in the 1990s (Shropshire Council, 2017). With limited development especially to the south and west, the mediaeval town by then sit at the southwest corner of the entire settlement.

5.2 Plan Unit Recognition

5.2.1 Historico-Geographical Analysis

Conzen (1988) has traced the formation and transformation of Ludlow from the early twelfth century until the Victorian period (1270–1830). His work mainly focuses on the Mediaeval town. A hierarchy of plan units was carefully produced (Fig. 5.2), disaggregating the walled area into different parts. As the mediaeval town has remained relatively consistent, plan unit division in the walled area across the three

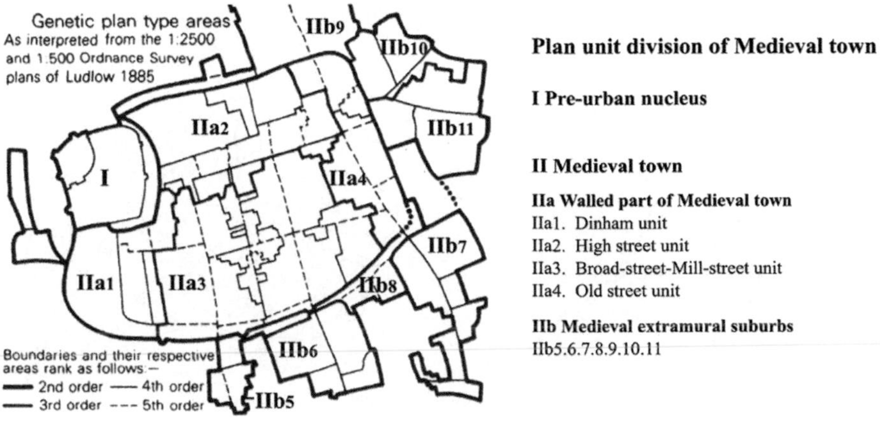

Fig. 5.2 Plan unit division of Mediaeval Ludlow, interpreted from 1885 Ordinance Survey Map (Conzen, 1988)

morphological periods has taken Conzen (1988)'s analysis for reference. Besides that, our study also investigates the subsequent development out of the town wall in the Victorian and Edwardian period (1831–1918) and the modern period (1919–2015). A hierarchy of plan units is identified based on the distinctive compositions of buildings, streets and plots in different periods (Fig. 5.3).

As shown in Fig. 5.3, the first-order plan units clearly distinguish areas that are generally developed within the same morphological period. That is, the castle (I), Mediaeval town (II), the Victorian and Edwardian development (III) and the modern development (IV). The second- order plan units identify the subzones that have homogenous built form character. For example, the Mediaeval town (II) is composed of the walled area (IIa), southern extensions (IIb) and northern extensions (IIc). The Victorian and Edwardian development (III) includes ribbon development along the different streets (IIId and IIIe). The modern development (IV) consists of inter-war area (IVg) and post-war area (IVh). The third-order plan units capture the finer-grained clusters of the interrelated and organized built form elements, usually seen as simple tissues. Examples are the grid street network within the walled area (IIa3), the bilateral ribbon development along the Corve Street (IIc9), continuous northern plot series as the extension of the north side of high-street market place (IIc10) and shortened and narrow plots along the Lower Galdeford (IIc11) in the Victorian and Edwardian period; the commercial and business development (IVh15), industrial sites (IVh16) and post-war residential settlements (IVh17) in the modern period.

A preliminary comparison of the plan units across the three morphological periods suggests that first- and second-order plan units, especially III and IV, are mostly structured by one to two main roads; and secondly, that third-order plan units are mostly characterized by distinctive street layouts, such as the gridiron pattern in the Mediaeval settlement (IIa3), the rectilinear pattern in the inter-war development (IVg14) and the cul-de-sac pattern (IVh17) in the post-war development. The

Medieval and early modern period

I The castle

II Medieval town

IIa Medieval town within the wall
IIa1 Dinham unit
IIa2 High Street unit
IIa3 Broad/Mill/Holgate Street unit
IIa4 Old Street unit

IIb Southern extensions
IIb5 Lower Mill/Broad Street unit
IIb6 Lower Holgate unit

IIc Northern extensions
IIc7 Corve Street unit
IIc8 Upper Galdeford Street unit
IIc9 Lower Galdeford Street unit

Victorian and Edwardian period

I The castle

II Medieval town

IIa Medieval town within the wall
IIa1 Dinham unit
IIa2 High Street unit
IIa3 Broad/Mill/Holgate Street unit
IIa4 Old Street unit

IIb Southern extensions
IIb5 Lower Mill Street unit
IIb6 Lower Broad Street unit
IIb7 Lower Holgate unit
IIb8 The outer fringe of town wall unit

IIc Northern extensions
IIc9 Corve Street unit
IIc10 Upper Galdeford Street unit
IIc11 Lower Galdeford Street unit

III Victorian and Edwardian development

IIId Gravel Hill/Julian Road unit

IIIe New Road unit
IIIe12 Union workhouse
IIIe13 Terraced housing

Modern period

I The castle

II Medieval town

IIa Medieval town within the wall
IIa1 Dinham unit
IIa2 High Street unit
IIa3 Broad/Mill/Holgate Street unit
IIa4 Old Street unit

IIb Southern extensions
IIb5 Lower Mill Street unit
IIb6 Lower Broad Street unit
IIb7 Lower Holgate unit
IIb8 The outer fringe of town wall unit

IIc Corve Street unit

IId Galdeford Street unit
IId9 Upper Galdeford Street unit
IId10 Lower Galdeford Street unit

III Victorian and Edwardian development

IIIe Gravel Hill/Julian Road unit

IIIf New Road unit
IIIf11 Union workhouse
IIIf12 Terraced housing

IV Modern development

IVg Inter-war development
IVg13 Henley Road unit
IVg14 Sandpits Road unit

IVh Post-war development
IVh15 Commerical and office development
IVh16 Industrial development
IVh17 Residential development
IVh18 Open space

Fig. 5.3 Plan unit divisions of Ludlow across three morphological periods (*Source* Authors)

following configurational analysis will be conducted to explore the extent to which the street network is related to the distribution of plan units.

5.2.2 *Configurational Analysis of Street Network*

The configurational analysis is conducted for every morphological period. Figures 5.4 and 5.5 show, respectively, the results of the global and local measurements. Descriptive statistics are also applied to further investigating the patterns of street configuration. The results of the global analysis uncover the most frequently travelled streets by trips between every single pair of street segments. They are shown as the long and continuous lines stretching from the mediaeval town to the periphery. These lines are generally thoroughfares of the city, hence are regarded as the integrators that connect the rest street segments which are accessible to various residential areas, such as High Street, Broad Street/Lower Broad Street, Corve/Old Street, Upper and Lower Galdeford streets.

The results of the local analysis at radius 400 m show the clustering pattern of the street network in both the town centre and the periphery. These clusters are generally

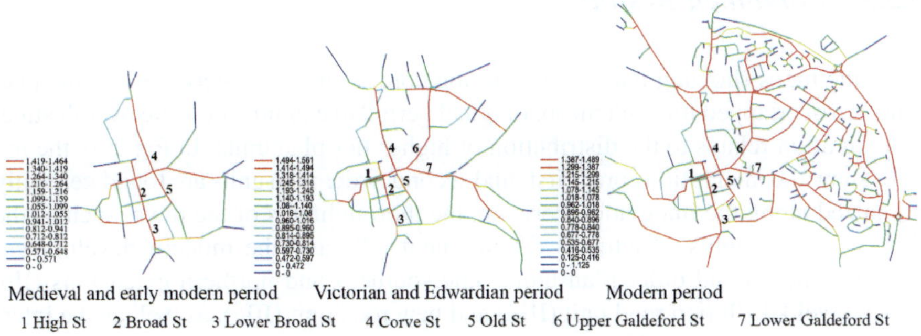

Fig. 5.4 Results of the Ludlow configurational analysis: the global measure of the street network of Ludlow in three morphological periods. Street segment lines are coloured from red to blue in accordance with their high-to-low configurational values of global NACH (*Source* Authors)

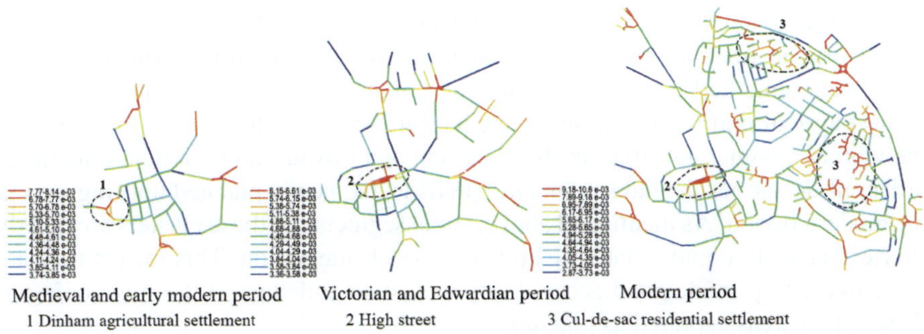

Fig. 5.5 Results of the Ludlow configurational analysis: the local measure of the street network of Ludlow at the radius of 400 m in three morphological periods. Street segment lines are coloured from red to blue in accordance with their high-to-low configurational values of local integration (*Source* Authors)

composed of higher density of shorter street segments intersecting with one another at right angles. As this market town is relatively small, the distance between origins and destinations is usually within a shorter walking distance. The local integration at radius 400 m could capture more clearly the local clustering pattern; thus, it is selected over the other radii.

The results of both the global and local measures are then investigated statistically, in order to more clearly capture the integration core[1] and the local cluster of street segments. A number of descriptive statistics, such as the percentile (10, 25 and 50%), and one standard deviation (σ) of the mean (μ), are applied. After repeated tests, it is found that the street segments with the top 20% of the global NACH values and the upper 50% of the local integration values could better describe the integration core and local cluster of the street segments of Ludlow across the three morphological periods.

5.2.3 Combined Results

First, the maps showing that the integration core of the street network and the plan units are overlapped for each morphological period to examine how the overall street configuration relates to the distribution of higher-tier plan units. In Fig. 5.6, the red lines show the integration core. First- and second-order plan units are found generally organized along the integration core, i.e. the thoroughfares of the town, such as the first-order plan units—Victorian development (III) and the modern development (IV) and the second-order plan units—the southern and northern extensions (IIb, IIc), Gravel Hill/Julian road unit (IIId) and new road unit (IIIe), as well as the inter-war development unit (IVg) and post-war development unit (IVh). One possible explanation for this observation is that accretion of a settlement usually tends to start with, and in turn reinforce, the extension of existing roads, especially the major ones.

Second, the maps highlighting the local cluster of the street segments are compared with the plan units to investigate how the clusters of street segments correspond to lower-tier plan units. Figure 5.6 shows that a number of the local clusters comprised of the orange lines correspond to plan units of the third order with distinctive socio, economic and cultural characteristics, such as the Dinham unit (IIa1)—the earliest agricultural settlement in the mediaeval period, the high-street unit (IIa2)—the vibrant commercial street in the Victorian and Edwardian period and the third-order cul-de-sac residential settlements (IVh14), the newly planned housing area in the modern period. As mentioned above, street segments of these clusters are mostly shorter ones and obtain relatively higher value of integration. This means they are more likely to provide good accessibility to sustain and support the functioning of these urban areas captured as plan units.

[1] Integration core is a pattern made of 10, 25% or 50% most integrating space (Hillier and Hanson, 1984).

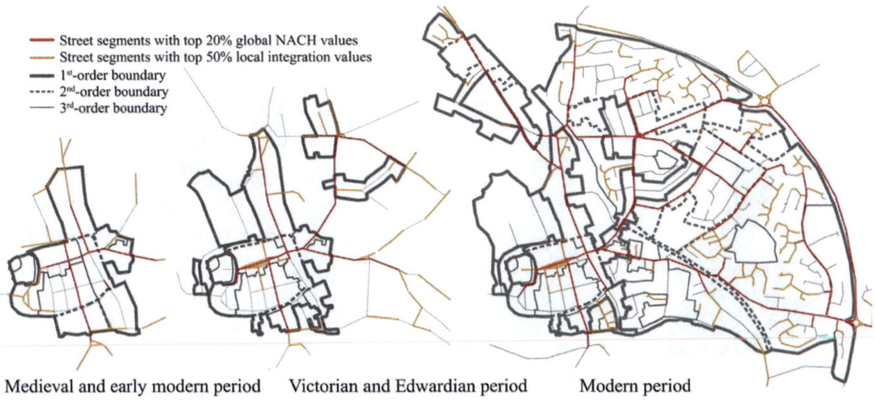

Fig. 5.6 Ludlow: the integrated approach in plan unit recognition. Patterns of the top 20% of the global NACH values (red) and the top 50% of the local integration values (orange) of the street segments overlapped with plan units across three morphological periods (*Source* Authors)

The results of the overlapped maps suggest that at least for this case study, the street configuration measured by the global NACH is correlated with the spatial distribution of plan units of the higher tier; the street configuration measured by the local integration largely corresponds to plan units of the lower tier. The formation of plan units follows the generative rule of urban space to a certain extent; thus, it can be argued that the configurational properties of the street network potentially offer an additional reference for plan unit recognition.

Third, the aggregate values of both the global NACH and local integration are calculated for plan units of all three tiers in each morphological period. Their pattern is explored and related to the historico-geographical process of the case study. Figure 5.7 shows that the global NACH values of first- and second-order plan units across three different periods. The newly developed plan units of the first order and second order in each period tend to gain lower global NACH values. The possible reason for the low-configurational value of the Victorian and Edwardian development (III) is that the move of the working-class households to the suburban areas under the Public Health Act 1875 (Conzen, 1988; Scott, 2013) made it less connected to the walled town than other plan units. As Ludlow continues to grow and expand the peripheries in the modern period, the Victorian and Edwardian development unit (III) became strategically important by connecting modern development unit (IV) with the town centre.

Figure 5.8 shows that the third-order plan units have distinctive local configurational values, corresponding to different types of development. In the modern period, the high-street unit (IIa2) is locally most integrated (5.16E−03), which is possibly related to its functioning as the business centre since the Mediaeval times that provides everyday amenities and shopping for local residents (Shropshire Council, 2017). This good accessibility in turn may also have positive feedback on maintaining and multiplying the business and retail functions of the high-street unit. This observation is also

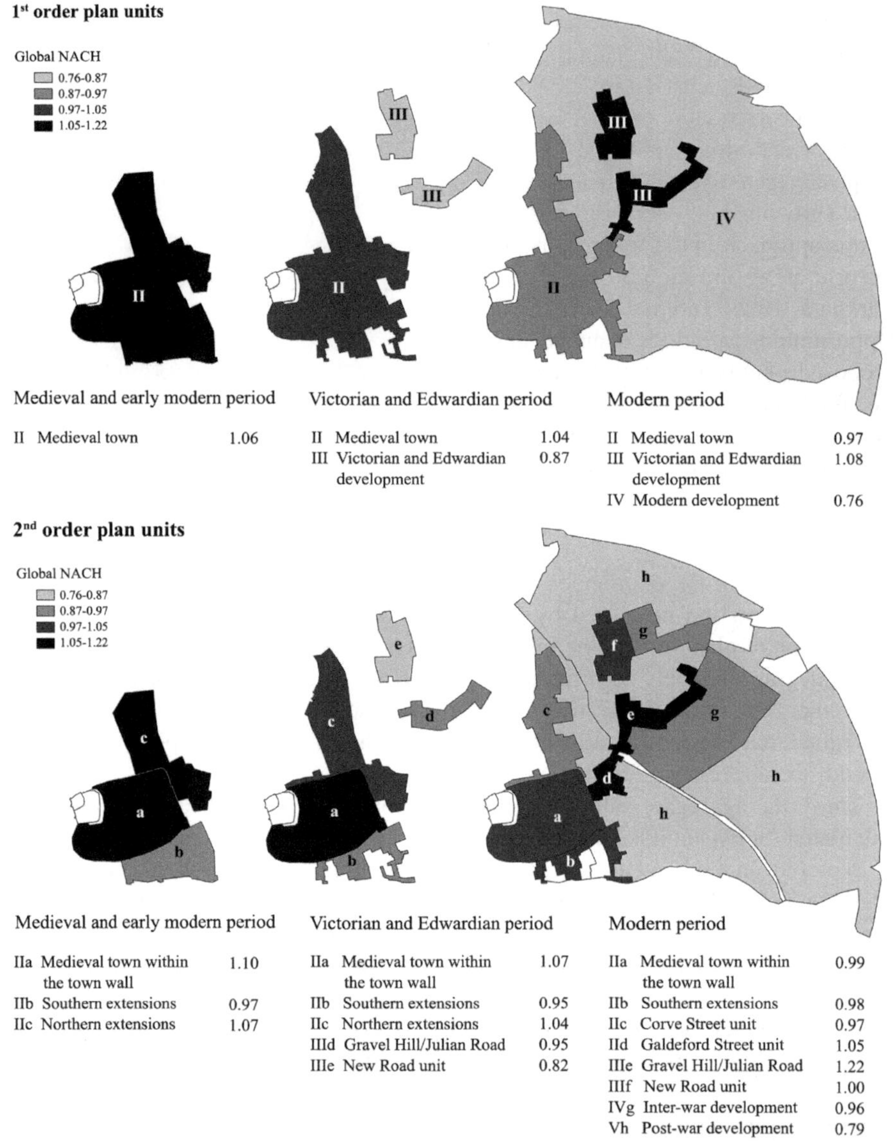

Fig. 5.7 Ludlow: the global NACH values of the first- and second-order plan units across three morphological periods (*Source* Authors)

consistent with that in the Victorian and Edwardian period. The cul-de-sac residential settlement (IVh17) is locally more integrated (4.52E−03) than that (3.96E−03) of the terraced housing unit (IIIf12). This is largely attributed to the advocated neighbourhood design since the 1960s, which aims to reduce the car traffic and increase local residents' outdoor activities (Crane, 2000). Yet the Broad Street/Mill Street/Holgate unit (IIa3) and Lower Broad Street unit (IIb6) originated from the Mediaeval period have similar configurational values (4.00E−03 and 4.06E−03 separately), which suggests that the street configuration alone may not fully recognize the lower-tier plan units.

The statistic results show that the historical process elaborated by the historico-geographical approach can, in turn, help interpret the configurational values of plan units. Based on that, the historico-geographical approach can also examine the role of the street configuration played in the continuity and transformation of plan units. It is found that plan units of the higher order (first and second) are more likely to be affected by the generic generative rules of urban space than plan units of the lower,

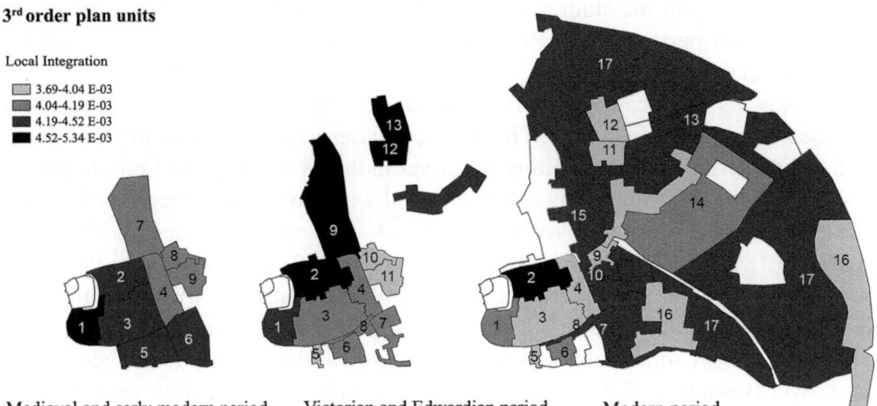

Medieval and early modern period

IIa1	Dinham unit	5.04
IIa2	High Street unit	4.27
IIa3	Borad/Mill Street unit	4.23
IIa4	Old Street unit	4.07
IIb5	Lower Mill/Broad Street	4.36
IIb6	Lower Holgate unit	4.21
IIc7	Corve Street	4.16
IIc8	Upper Galdeford Street	4.19
IIc9	Lower Galdeford Street	4.17

Victorian and Edwardian period

IIa1	Dinham unit	4.22
IIa2	High Street unit	5.34
IIa3	Borad/Mill Street unit	4.06
IIa4	Old Street unit	4.07
IIb5	Lower Mill Street unit	3.69
IIb6	Lower Broad Street unit	4.08
IIb7	Lower Holgate unit	4.19
IIb8	Outer fringe of town wall	4.22
IIc9	Corve Street	4.91
IIc10	Upper Galdeford Street	4.02
IIc11	Lower Galdeford Street	3.89
IIIe12	Union workhouse	4.81
IIIe13	Terraced housing	4.92

Modern period

IIa1	Dinham unit	4.13
IIa2	High Street unit	5.16
IIa3	Borad/Mill Street unit	4.00
IIa4	Old Street unit	3.83
IIb5	Lower Mill Street unit	3.70
IIb6	Lower Broad Street unit	4.06
IIb7	Lower Holgate unit	4.31
IIb8	Outer fringe of town wall	4.39
IId9	Upper Galdeford Street	3.99
IIId10	Lower Galdeford Street	4.40
IIIf11	Union workhouse	3.90
IIIf12	Terraced housing	3.96
IVg13	Henley Road unit	4.31
IVg14	Sandpits Road unit	4.17
IVh15	Commerical and office development	4.45
IVh16	Industrial development	4.04
IVh17	Residential development	4.52

Fig. 5.8 Ludlow: the local integration values of the third-order plan units across three morphological periods (*Source* Authors)

as many other factors, such as development guidelines, may influence the formation of plan units in the local scale.

The above discussions have systematically correlated the results of the historico-geographical and configurational approaches through overlapped maps and descriptive statistics. The case study of Ludlow has demonstrated the complementary use of both approaches. The findings show that the historical transformation of Ludlow is not only influenced by the external social, economic and cultural factors, but also by the generative rules of urban space. How this framework works in a completely different urban context will be explored next.

5.3 The Case Study: Chinatown, Singapore

Chinatown is a historic district located on the south bank of Singapore River in the Central Region of Singapore. Three morphological periods of Chinatown are recognized based on the studies of Liu et al. (2001) and Kong (2011), that is, the British Colonial period (1822–1920), the transitional period (1920–1965) and the cosmopolitan period (1966–2015).

Chinatown's physical development began in 1822 when it was allocated for Chinese settlement (Liu, 1995). This ethnic quarter extended westward and northward to New Bridge Road and Singapore river in the 1840s (Fig. 5.9 Left). Land was laid out in standard gridiron plan characterized with narrow and deep lots for shophouses along the Singapore River and the sea. The east coastline was expanded in the 1870s and 1890s through land reclamation (Chua, 1989). It was until the 1920s that the street system was well developed, along with the southern expansions along the Neil Road and Tg Pagar Road. The 1950s and 1960s were relatively stable periods for Chinatown except some notable developments in Raffles Place and the newly built Telok Ayer Basin (Fig. 5.9 middle). From the mid-1950s, the Central Area of Singapore was targeted as a central business and financial district (CBD) in Singapore's planning concepts (1958 and 1971 Concept Plan), and during the 1970s and 1980s, there was large-scale demolition and reconstruction of the historic area (Kong, 2011). The previous fine-scale shophouses and small alleyways were demolished and made way for new development composed of high-rise office buildings. The CBD area of about 35 ha was named as 'Golden Shoe' for its high-land price and shoe-like shape (Chua, 1989). Constrained by Singapore River on the north and conservation areas (including gazetted Kreta Ayer, Telok Ayer, Bukit Pasoh and Tanjong Pagar) on the west, the Golden Shoe area began to spill over until it reached Keppel Road southward and Cantonment Road and Tg Pagar Road westward. In the 1990s when the Golden Shoe area's westward expansion has reached its limits, it began to expand seaward to Marina south - the newly reclaimed land (Fig. 5.9 right).

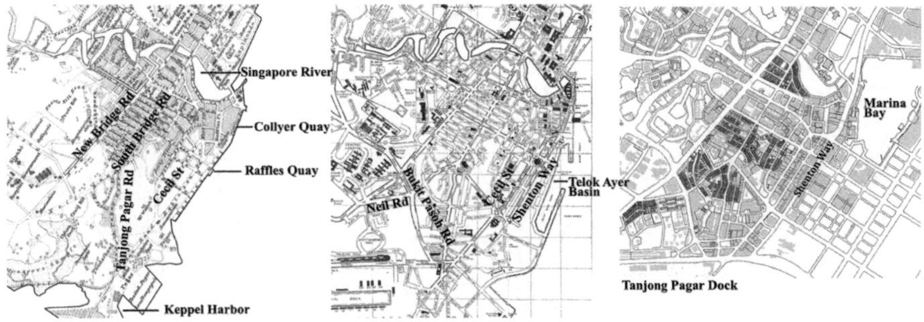

Fig. 5.9 Left. Plan of Chinatown in 1900 (*Source* German Encyclopaedia). Middle. Plan of Chinatown in 1950 (*Source* Singapore's online OneMap). Right. Cadastral map of Chinatown in 2015 (*Source* Singapore Land Authority)

5.4 Plan Unit Recognition

5.4.1 Historico-Geographical Analysis

The plan units across the three morphological periods are presented in Fig. 5.10.

The first-order plan units remain relatively consistent in the first two morphological periods. In the transitional period (1921–1965), two more plan units (IV and V) grow to the east and west of the previous Chinese settlement. When it comes to the cosmopolitan period (1966–2015), there is a lack of continuity in the plan units due to the large-scale demolition and construction in this period. The remaining traditional settlement is divided into four first-order plan units: Boat Quay unit (I), Chinatown unit (II), Bukit Pasoh—Tanjong Pagar unit (IV) and Blair Rd unit (V). Two new plan units of first order—Maxwell Road unit (III) and Urban Redevelopment unit since 1970s (VI)—emerged under the amalgamation of the previous plots and reconstruction of new buildings.

The second-order plan units are subzones with homogenous built form character, which did not change much in the first two morphological periods. In the cosmopolitan period (1966–2015), several new plan units of the second-order appear, such as the post-1970s urban redevelopment unit (VI), Golden shoe unit (VIf), southern extension of Golden shoe unit (VIg) and western extension of Golden shoe unit (VIh).

In comparison with Ludlow, the number of tiers in the hierarchy of plan units in Chinatown is fewer. As Chinatown is relatively smaller in size and shorter in history, the lower degree of historical stratification makes it difficult to further delimit even smaller urban areas.

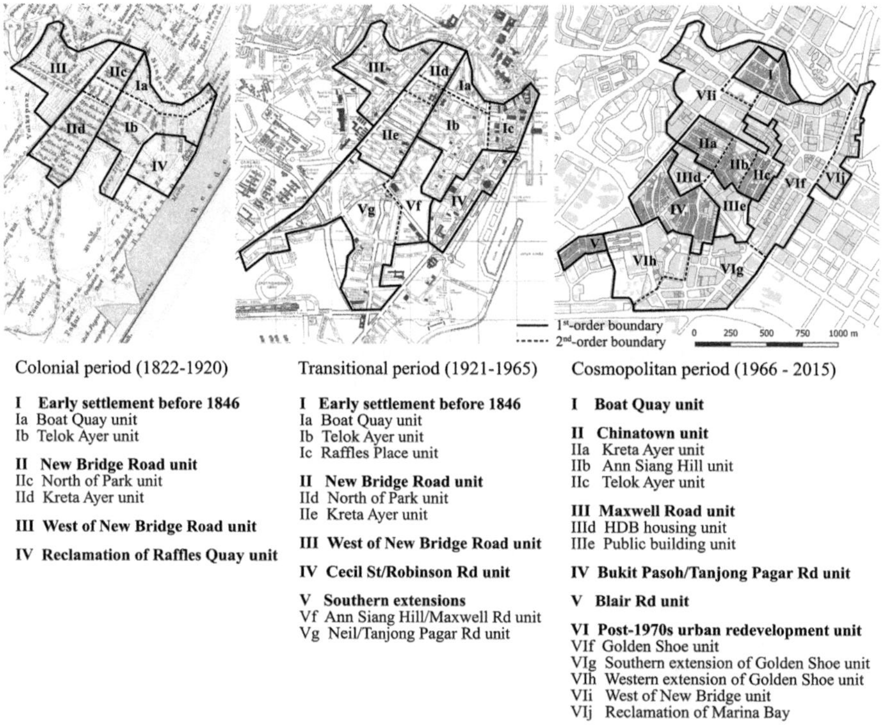

Fig. 5.10 Plan unit divisions of Chinatown across three morphological periods (*Source* Authors)

5.4.2 Configurational Analysis of Street Network

The configurational analysis for Chinatown is conducted for each morphological period (Figs. 5.11 and 5.12). Descriptive statistics are also applied to further investigating the patterns of street configuration.

The results of the global analysis highlight the integration core in Fig. 5.11, which is consisted of the most frequently travelled streets by trips between every single pair of street segments. Consistent with Ludlow's analysis, they are shown as the long and continuous street segments which are usually the thoroughfares in the city, such as New Bridge Road, South Bridge Road, Upper Pickering Street and Neil Road.

The results of the local analysis capture the local clusters of the street segments at radius 800 m in Fig. 5.12. They are made up of high density of shorter street segments. Local integration at radius 800 m could more clearly depict Chinatown's local clustering pattern than the other radii.

The results of both the global and local measures are investigated statistically, in order to more clearly capture the integration core and the local cluster of street segments of Chinatown. As with Ludlow, the top 20% of global NACH values and

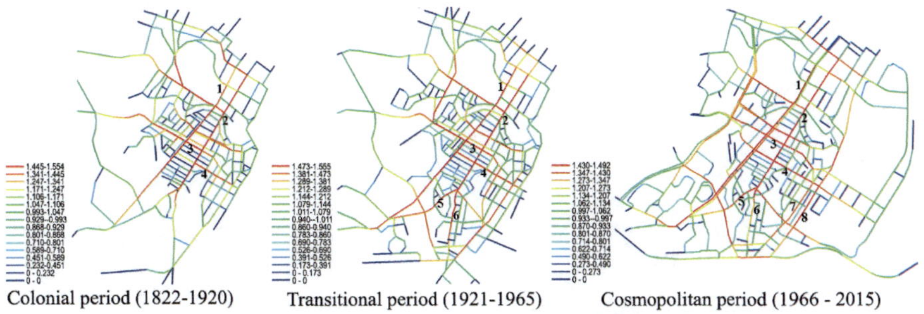

Fig. 5.11 Results of the Chinatown configurational analysis: the global measure of the street network of Chinatown in three morphological periods. Street segment lines are coloured from red to blue in accordance with their high-to-low configurational values of global NACH (*Source* Authors)

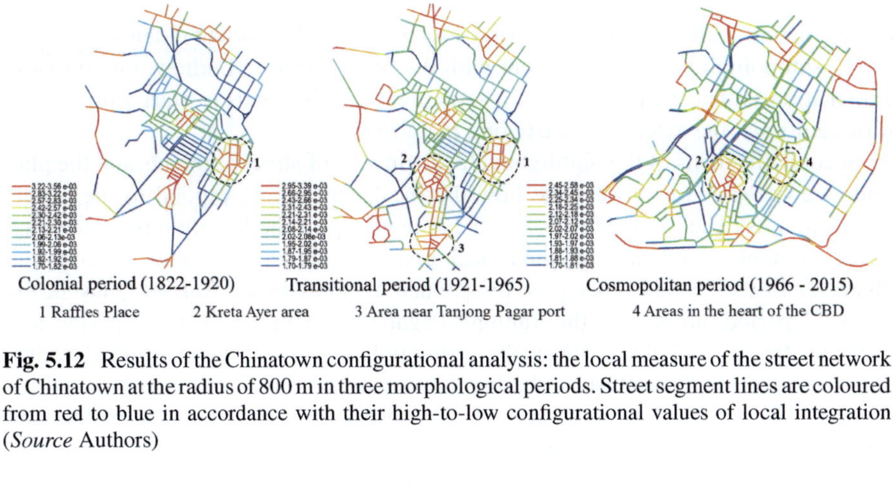

Fig. 5.12 Results of the Chinatown configurational analysis: the local measure of the street network of Chinatown at the radius of 800 m in three morphological periods. Street segment lines are coloured from red to blue in accordance with their high-to-low configurational values of local integration (*Source* Authors)

the top 50% of local integration values could better highlight the global and local street network structures across the three morphological periods.

5.4.3 Combined Results

First, the maps showing the integration core and the plan units are overlapped for each morphological period to explore how the overall street configuration relates to the higher-tier plan units. The red lines in Fig. 5.13 represent the integration core. The first-order plan units are found locating along the integration core which is mostly the thoroughfares of the city, such as the Cecil St and Robinson Rd unit and Southern

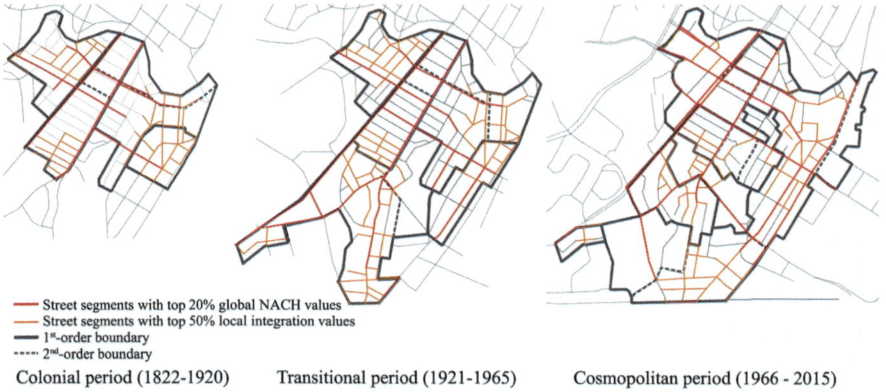

Fig. 5.13 Chinatown: the integrated approach in plan unit recognition. Patterns of the top 20% global NACH values (red) and 50% local integration values (orange) of street segments overlapped with plan units across three morphological periods (*Source* Authors)

extension unit in the Transitional period, and the post-1970s urban redevelopment in the Cosmopolitan period. Consistent with the observation in Ludlow, one possible explanation is that the accretion of a settlement usually tends to start with, and in turn, reinforce, the extension of existing main roads.

Second, the maps of the highlighted local clusters of street segments and the plan units are compared to investigate how the local clusters correspond to the lower-tier plan units. The orange lines (Fig. 5.13) made up of the locally most integrated street segments could capture some local areas with distinguished socio-economic character in the second-order plan units, such as areas in the Raffles Quay in the Colonial period, areas near the Tanjong Pagar port in the transitional period and areas in the heart of the CBD in the cosmopolitan period. As mentioned above, street segments of these clusters obtain relatively higher values of integration, which are more likely to provide good accessibility to sustain and support the trading and business functions of these urban areas.

The results of the overlapped maps show that the distribution of first-order plan units is related to the street configuration measured by the global NACH. In contrast, a number of the lower-tier plan units in Chinatown are not captured by the locally integrated street segments, such as the public building unit (IIIe) and western extension of GS unit (VIh). The underlying reason is that Chinatown has a dramatically different urban context from Ludlow, as it has undergone large-scale urban redevelopment. Some intrinsic spatial structures of urban areas are likely to be associated with historical urban process while others are not. This indicates that the street configuration may play a weaker role in providing an additional reference for plan unit recognition in Chinatown than Ludlow.

Third, the aggregate values of both the global NACH and local integration are measured for plan units of all three tiers in each morphological period. Their patterns are explored and related to the historical urban process of Chinatown. There are two observations in Fig. 5.14: the first is that the newly developed plan unit tends to have lower global NACH values than that of older plan units. For example, Cecil and Robinson Rd unit (IV) in the Transitional period has the lowest configurational value (1.02). This could be explained by the fact that areas along Shenton Way within this plan unit remained mostly vacant and undeveloped at that time (Chua, 1989). The second observation is that when a new plan unit replaces the previous urban fabric, its global NACH values tend to be higher than the older plan units. Take the post-1970s urban redevelopment unit (VI) in the cosmopolitan period for instance, its higher configurational value (1.04) is possibly related to its designation as the Central Business District (CBD) of Singapore since the 1970s (Dale, 1999). Further, this high spatial accessibility may in turn have positive feedback on the CBD's financial and commercial functioning.

Figure 5.14 also shows that the second-order plan units are characterized with distinctive local configurational values. For example, in the Cosmopolitan period, the newly developed plan units, such as HDB housing unit (IIId), Golden shoe unit (VIf) and Southern extension of Golden shoe unit (VIg), exhibit high-local integration values. The underlying reason is that given the limited land space and high-land values in the downtown core of Singapore, urban redevelopment mainly takes shape in smaller urban blocks, seeking potentials to increase density and accommodate more diverse enterprises and businesses by the joint efforts of the public and private sectors (Chua, 1989; Dale, 1999). Yet the Kreta Ayer unit (IIa) and AnnSiang Hill unit (IIb) have similar configurational values (2.03E−03 and 2.04E−03), which also suggests that the street configuration alone may not fully recognize lower-tier plan units.

The value comparisons show that the historical urban process elaborated by the hisotrico-geographical approach can help interpret the configurational values of plan units to a certain extent. As with the observation in Ludlow case study, plan units of the first order are more likely to be affected by the generic generative rules of urban space than plan units of the second order, as many other factors may influence the formation of plan units in the local scale. The possible reasons are explained next.

5.5 Discussion

On the one hand, the configurational approach could uncover the intrinsic spatial structures of urban areas that are likely to be associated with the historical urban process. In both case studies, the integration-segregation pattern of the street network is related to the spatial distribution of plan units. The higher-tier plan units in both cases are found along the integration cores which are mostly the thoroughfares of the city. The lower-tier plan units could be captured by local clusters of the street segments. The slight difference is that the local integration measure in Ludlow could correspond to more plan units of the lower tier than that in Chinatown. As Ludlow grows incrementally outward from the existing periphery of the mediaeval town, urban settlement in each morphological period exhibits different street layouts. Thus,

1st order plan units

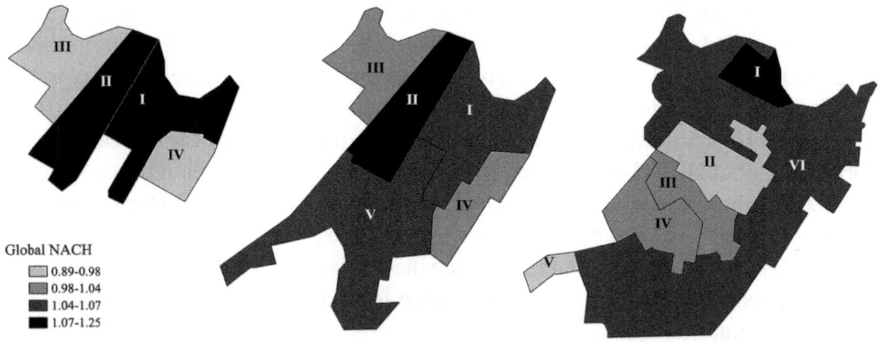

Colonial period	
I Early settlement before 1846	1.10
II New Bridge Road unit	1.19
III West of New Bridge Road	0.94
IV Reclamation of Raffles Quay	0.94

Transitional period	
I Early settlement before 1846	1.06
II New Bridge Road unit	1.25
III West of New Bridge Road	1.04
IV Cecil St/Robinson Rd unit	1.02
V Southern extensions	1.05

Cosmopolitan period	
I Boat Quay unit	1.07
II Chinatown unit	0.97
III Maxwell Rd unit	0.99
IV Bukit Pasoh/Tanjong Pagar Rd unit	1.03
V Blair Rd unit	0.89
VI Post-1970s urban redevelopment unit	1.04

2nd order plan units

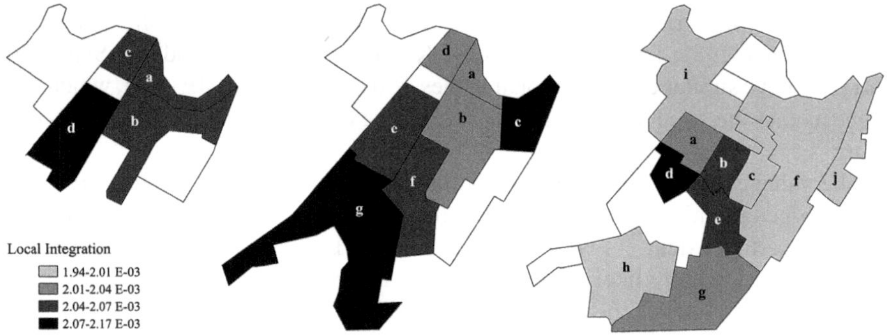

Colonial period	
Ia Boat Quay unit	2.06
Ib Telok Ayer unit	2.07
IIc North of Park unit	2.07
IId Kreta Ayer unit	2.14

Transitional period	
Ia Boat Quay unit	2.02
Ib Telok Ayer unit	2.02
Ic Raffles Place unit	2.13
IId North of Park unit	2.01
IIe Kreta Ayer unit	2.07
Vf Ann Siang Hill/Maxwell Rd unit	2.05
Vg Neil Rd/Tanjong Pagar Rd unit	2.14

Cosmopolitan period	
IIa Kreta Ayer unit	2.03
IIb Ann Siang Hill unit	2.04
IIc Telok Ayer unit	1.98
IIId HDB housing unit	2.17
IIIe Public building unit	2.04
VIf Golden Shoe unit (GS)	2.01
VIg Southern extension of GS	2.02
VIh Western extension of GS	1.94
VIi West of New Bridge unit	1.98
VIj Reclamation of Marina Bay	2.00

Fig. 5.14 Chinatown: the global NACH values of first-order plan units and the local integration values of second-order plan units across three morphological periods (*Source* Authors)

street configuration alone could capture lower-tier plan units. Yet Chinatown underwent large-scale urban redevelopment, which has a major impact on the compositions of plots and building blocks. These two form elements in turn characterize urban settlement in each morphological period rather than the planned urban grid that has not changed much. Therefore, street configuration plays a relatively limited role in lower-tier plan unit recognition. From the above discussion, it can be argued that besides cartographic sources, plot-by-plot survey and documentary evidence, the configurational approach has the potential to provide an additional reference to plan unit recognition in an objective manner, and its usefulness is dependent on the specific street network patterns.

On the other hand, the historical urban process by the historico-geographical approach in both case studies could help explain the changes in the configurational values of different plan units to a certain extent. Both case studies show that the newly developed plan unit tends to have lower global NACH values than that of the older plan unit due to the outward urban growth and expansion. Yet in both Ludlow and Chinatown, not all local settlements can be recognized by the configurational patterns or values. Other factors, such as development strategies and planning regulations and guidelines, may outweigh the impact of street configuration in the transformation of plan units.

It can therefore be seen that the combination of the historico-geographical and configurational approaches is capable of providing enhanced understandings of the historical transformations of urban form. It is conditioned by the external social-economic factors; at the same time, it is also underscored by the more generic generative rules of urban space. The extent to which these two aspects influence the transformation of urban form is dependent on the different development process of urban areas—incremental growth or fast-changing urban landscapes.

This chapter has proposed a methodological framework that combines the historico-geographical and configurational approaches to understand the historical transformation of urban form by two case studies. This integrated framework can be seen as a fundamental step to addressing the major challenge of coordinating and combining different morphological schools that faces the field of urban morphology today. Building on it, the next chapter will apply the integrated framework to evaluate the strength and weakness of landscape units based on their current historical expressiveness and spatial potential. In addition, this integrated framework has the potential to address some prevalent issues in urban landscape management, such as the ill-delimited geographical boundaries of historic urban areas especially in developing countries (Whitehand, 2007; Whitehand & Gu, 2007), which will be discussed in Part III.

References

Chua, B. H. (1989). *The golden shoe: Building Singapore's financial district*. Urban Redevelopment Authority.

Crane, R. (2000). The Influence of Urban Form on Travel: An Interpretive Review. Journal of Planning Literature, 15(1), 3–23.

Conzen, M. R. G. (1988). Morphogenesis, morphological regions and secular human agency in the historic townscape, as exemplified by Ludlow. *Urban historical geography: recent progress in Britain and Germany*, 253–272.

Dale, O. J. (1999). *Urban planning in Singapore: The transformation of a city*. Oxford University Press.

Hillier, B., & Hanson, J. (1984). The social logic of space. New York: Cambridge University Press.

Liu, G. (1995). *Pastel portraits: Singapore's Architectural Heritage*. Singapore Coordinating Committee.

Liu, G. (2001). Singapore: a pictorial history 1819–2000. London: Curzon.

Kong, L. (2011). Conserving the past, creating the future: Urban heritage in Singapore. Singapore: Urban Redevelopment Authority.

Scott, P. (2013). *The making of the modern British home: the suburban semi and family life between the wars*. OUP Oxford.

Shropshire council. (2017). *Ludlow market town profile*. Retrieved from UK: https://shropshire.gov.uk/media/20655/market-town-profile-ludlow-ev07310.pdf

Whitehand, J. W. R. (2001). British urban morphology: The Conzenion tradition. *Urban Morphology, 5*(2), 103–109.

Whitehand, J. W. R. (2007). *Conzenian urban morphology and urban landscapes*. Paper presented at the Proceedings, 6th International Space Syntax Symposium, Istanbul.

Whitehand, J. W. R., & Gu, K. (2007). Urban conservation in China: Historical development, current practice and morphological approach. *Town Planning Review, 78*(5), 643–670.

Chapter 6
Integrated Examination of Urban Form: Historicity and Socio-economic Vibrancy

Abstract This chapter presents the selected Chinese city—Nanjing as a case study to demonstrate the integrated methodological framework to evaluate the character and strength of urban form in terms of its historicity and spatial potentials for generating socio-economic vibrancy. The historical quarter of Nanjing—Old Chengnan District is mainly focused on. Through combining the analytical results of landscape units and spatial potentials, this framework is able to identify not only which parts of landscape units exhibit both distinguished historic character and good spatial potential that have comparative advantage for conservation, and which parts confront deteriorated historical expressiveness and poor spatial potential that may need physical improvement for future redevelopment. In addition, it could also provide an understanding of how different landscape units are spatially and socially connected, and how changes to one may affect another. This integrated framework provides a finer-grain and cross-scale evaluation of landscape units, which has implications for the management of historical urban landscape in view of what to change and what to maintain.

Keywords Historico-geographical approach · Landscape unit · Configurational approach · Place syntax · Functional mix · Nanjing

6.1 The Case Study: Nanjing, China

Since the economic reform of 1978, an unprecedented urbanization of cities has been taking place. The historic inner city has become the focus of urban development and regeneration processes, which often lead to conflicts over the preservation of heritage values. Large-scale demolition of historic urban fabric was prevalent. New urban and architectural forms have been developed at speed, but with little or no reference to existing urban context. A large number of cities in China are currently in such as status, in particular the listed Historic and Cultural Cities. In recent years, it has reached an agreement that a careful and sensitive planning and management of the economic resource so as not be in conflict with the concern to protect and conserve the historic values (Fang et al., 2012). Yet without a sound theoretical basis, the effective strategies for deciding what to conserve and how to conserve it are still

X. Li and Y. Zhang, *Conserving and Managing Historical Urban Landscape*, The Urban Book Series, https://doi.org/10.1007/978-981-19-4222-8_6

lacking. Therefore, Nanjing was selected for this study as a typical Chinese historic city which urgently requires conceptual framework and analytical approach to assess its historical expressiveness and spatial potentials for generating socio-economic vibrancy.

Before zooming in on the Old Chengnan District, a brief introduction of Nanjing is put below. Nanjing was listed in the first batch of the Chinese Precious Historic and Cultural cities in the 1980s. It is a city with a history of more than 2500 years, sited in the Yangtze River Delta region, about 300 km north-west of Shanghai. It is an important administrative centre and provider of public services in Jiangsu Province. It has the world's longest surviving city wall, and the area within the Ming city wall is 43 km^2.

Nanjing has a layered history; different periods have left distinctive residues in the city. It used to be the capital of China in the six dynasties of Wu, Dongjin, Song, Qi, Liang and Chen (229–280 AD; 317–419, 420–589 AD), South Tang Dynasty (937–975 AD), Ming Dynasty (1368–1420 AD) and the Republican period (1911–1937). Figure 6.1 (left) shows that in the South Tang Dynasty, the imperial palace (in dark grey) moved southward and the city wall of around 14 km was built to enclose the palace and urban settlement along the River Neiqinhuai. In the Ming Dynasty, an even larger city wall was built, which included the areas in the far north, the newly built imperial palace (in dark blue) on the east and the previous South Tang settlement. The enclosed city size doubled as well, with a perimeter of around 35 km (Xue, 2015). The residential and commercial areas were developed mainly along the River Neiqinhuai, as the waterway was important transporting goods during the Ming and Qing dynasties (Han, 2004). Figure 6.1 (right) shows that the street networks of the walled city before 1911 were formed largely under the influence of waterways and previous imperial palaces. For example, Dagongfang Street and Yudao Street as the north–south axis extended southward to the wall gates. The street networks were further developed during the Republican period. Axial streets, such as Zhongshan Rd, Zhongyang Road and Zhongzheng Road, were built under the Capital plan published in 1929 and have become the most important streets in the city.

There are three historic districts in the walled area of Nanjing: Ming Palace District (6.5 km^2), Old Chengnan District (6.9 km^2), and Drum Tower District (6.2 km^2) (Fig. 6.2). Old Chengnan District is defined by the city wall and Shengzhou-Jiankang Road. Its streets patterns, plots and building block plans were developed under the strong influence of the River Neiqinhuai and the Ming city wall, dating back to the Ming and Qing dynasties. It is now characterized with a juxtaposition of settlements ranging from different historical periods, such as Ming and Qing residential settlement, post-1949 industrial areas, 1990s multi-storey housing and recent large-scale redevelopment (Fig. 6.3). For its rich historical layers, Old Chengnan District is selected as the case study to test the analytical framework that combines the historico-geographical approach and the configurational approach.

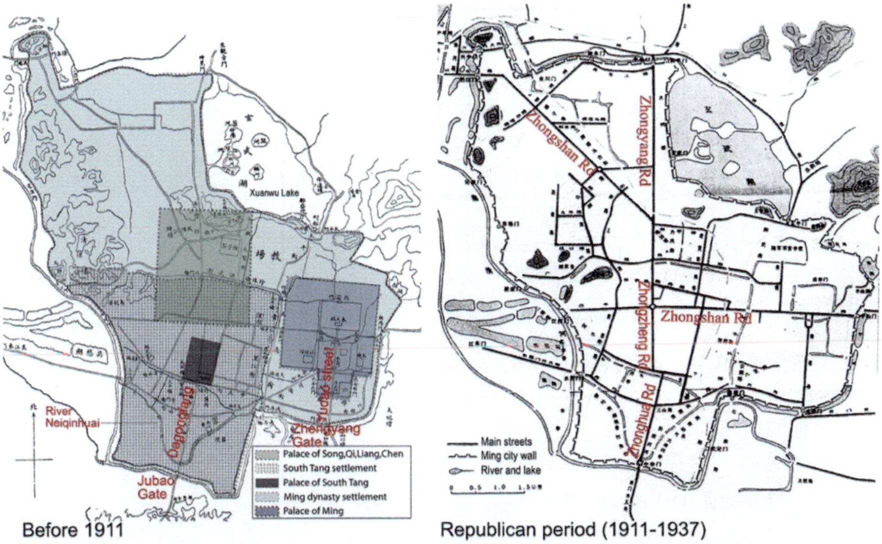

Fig. 6.1 Right: transformation of Nanjing as capitals throughout history. Left: street networks in the walled city around 1911 (Chen, 2017; Nanjing Local chronicle committee, 2008)

Fig. 6.2 Bird's eye view of Old Chengnan District in the walled area from the south (*source* http://uav.news.cn)

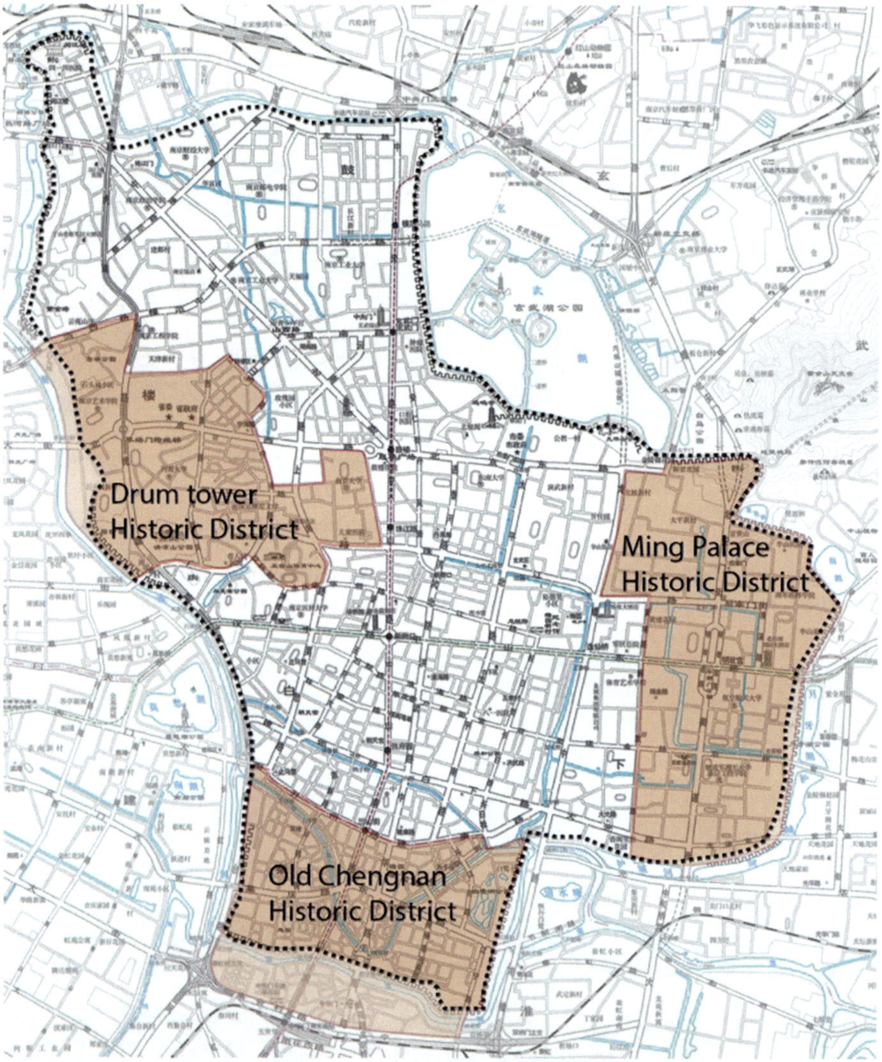

Fig. 6.3 Three historic districts in the old city of Nanjing in 2010 (*source* authors)

6.2 Plan Unit Recognition

This section combines the historico-geographical and configurational approaches to explore how spatial configuration could contribute to plan unit recognition in a different context to the previous case study, and to what extent the generative rules of space are related to plan unit formation and transformation. Sections 6.2.1 and 6.2.2 explain how the plan unit analysis and street network structure analysis were carried out separately. In Sect. 6.2.3, the combined analysis is presented to answer these questions.

6.2.1 Historico-geographical Analysis

A number of researchers (Chen & Thwaites, 2013; Xue, 2015; Ye et al., 2014) have come up with a consensus on the division of Nanjing's urban development periods, which are 1865–1911, 1912–1949, 1949–1978, 1979–1998. The year 1865 marks the beginning of Nanjing's modernization and industrialization, when Jinling Military Arsenal was built by Li Hongzhang during the Self-Strengthening Movement, followed by the first modern road—Jiangning Road—built up in 1899 by Zhang Zhidong (Xue, 2015). 1912 marks the beginning of the Republican period in Nanjing, when stable economic growth and government projects began to take place. During the Republican period, the Capital Plan (*shoudu jihua*) of 1929, the first planning document of Nanjing, greatly impacted the city with regard to functional zoning and transportation planning, though only part of it was implemented (Xue, 2015).

After the establishment of the People's Republic of China in 1949, Nanjing continued to be an important economic and political centre even after its loss of capital status (Chen & Thwaites, 2013; Xue, 2015). The land use system transformed from privately owned to state-owned, and land was administratively allocated by the central government. The state-led industrialization strategies of 'turning cities from consumption to production centre' (Xie & Costa, 1991, p. 280) prioritized the construction of large- and medium-sized industrial enterprises in the old city and at the fringe of the Ming city wall. Meanwhile, work-units (*danweis*) provided as welfare housing for urban workers mushroomed in the 1950s. Since economic reform was initiated in 1978, a series of institutional and financial policies have been erected, such as the legislation of transferable land use right in 1988. This reform has led to the emergence of an urban land and real estate market. As a result, urban constructions speeded up and grew outside the Ming city wall. The built-up area increased from 32 km^2 in 1947 to 144 km^2 in 1997 (He & Cui, 2000). The change of housing system in 1998 from welfare housing provision to commodified housing pushed Nanjing into another period of rapid development and redevelopment (Yang & Chen, 2014). Various types of housing, high-rise offices, hotels and shopping malls have reshaped the urban fabric and skylines of Nanjing.

Based on the above discussions, in this study four morphological periods are identified for plan unit analysis: (1) early modern and modern period (1865–1948); (2) planned economy period (1949–1978); (3) transition period (1979–1997); (4) market economy period (1998–2015).

Nanjing's rich historical resources provide a basis for detailed historico-geographical analysis. A time series of cartographic data, such as aerial map, cadastral map and land use map, were imported into ArcGIS. By georeferencing them to Nanjing's 2015 Google Satellite map (in WGS 84 coordinates), the rigor of the original maps in different scales could be enhanced. These maps were compared visually with the aid of transparent overlays in ArcGIS, which provides a direct view of urban form evolution over a long period of time.

Early modern and modern period (1865–1948)

During this morphological period, axial streets, such as Zhonghua Street and Shengzhou-Jiankang Street, were built under the Capital plan published in 1929 and became the most important streets in the city. The street alleyways in the residential settlement were formed under the strong influence of the River Neiqinhuai.

The majority urban fabric was homogeneous, consisted of long and narrow plots, courtyard buildings and street alleyways. The land ownership at that time was private. Land was subdivided by landowners into standardized rectangular parcels for sale or occupation. Figure 6.4 shows that a standardized building code was erected by the central government that an ordinary residential building shall have 'three bays and five rafters'.

Along the River Neiqinhuai plots were large in size with fewer road crossings; these housed mainly institutions and schools, developed from previous clan and religious building blocks (*Citang*). At the fringe of the Ming city wall, there were also larger, irregular plots with low building coverage; these mainly consisted of

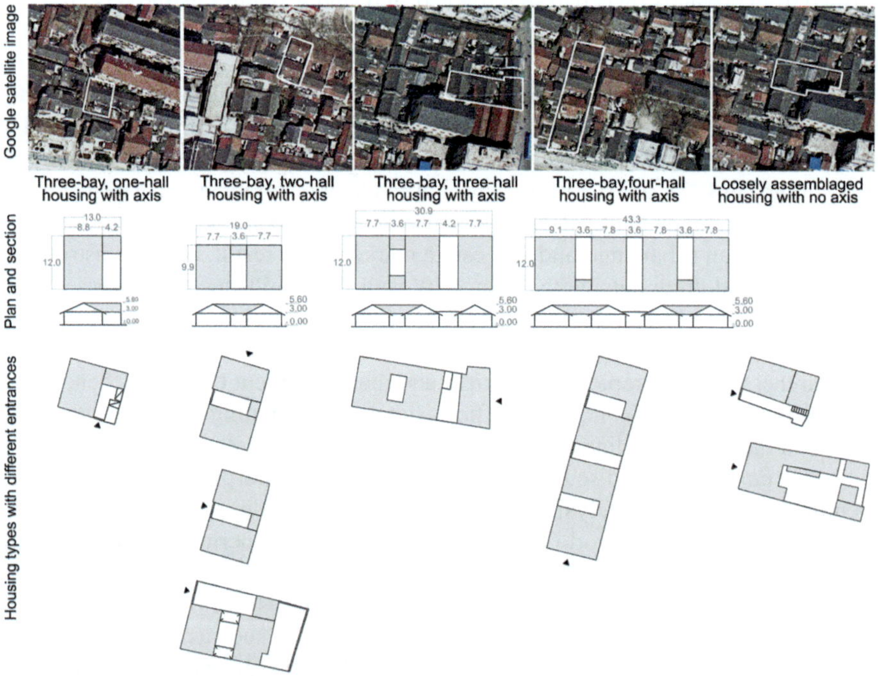

Fig. 6.4 Typologies of courtyard housing (*source* authors). They are 'three-bay' in width with a varied number of 'halls' in depth. The width of building and plot remains consistent based on the standardized building code, while the depth of plot could reach up to 60 m. They are usually one or two-storey high, arranged in symmetric layout. Small courtyards are designed to allow ventilation and sunlight to come in

Fig. 6.5 Plan unit division in the early modern and modern period (1865–1949) (base map: the 1937 cadastral map)

industrial sites, agricultural land, religious sites and vacant land, such as IIId30, IIIf 39, IIIh42 (Fig. 6.5).

Figure 6.6 shows that the first-order boundaries identify three plan units. The plan unit of Meander of the River Neiqinhuai (I) is consistent with Whitehand and Gu (2017)'s definition of an inner fringe belt. It is hypothesized that back to the earliest days, the River Neiqinhuai, as a defensive feature, may have served as a strong fixation line, restricting the outward residential growth. Zones of large plots with lower building density tended to form along the waterway. Similarly, the plan unit of inner fringe of the Ming city wall (III) is consistent with Whitehand and Gu (2017)'s definition of a Ming fringe belt. The Ming city wall was built up in the fourteenth century for military defence. It also served as a strong fixation line, restricting the outward residential growth. These two plan units survived through Nanjing's long feudal periods well into the twentieth century. The second-order boundaries are influenced by the clusters of different building blocks, such as Ia4 (Zhonghua primary school) and Ib11 (Nanjing hospital) within the Meander of the River Neiqinhuai (I), IIId30 (Nanjing machine plant) and IIIh42 (vacant land) within the inner fringe of the Ming city wall (III). The third-order boundaries distinguish individual establishments, such as government institutions Ia1 and Ia2.

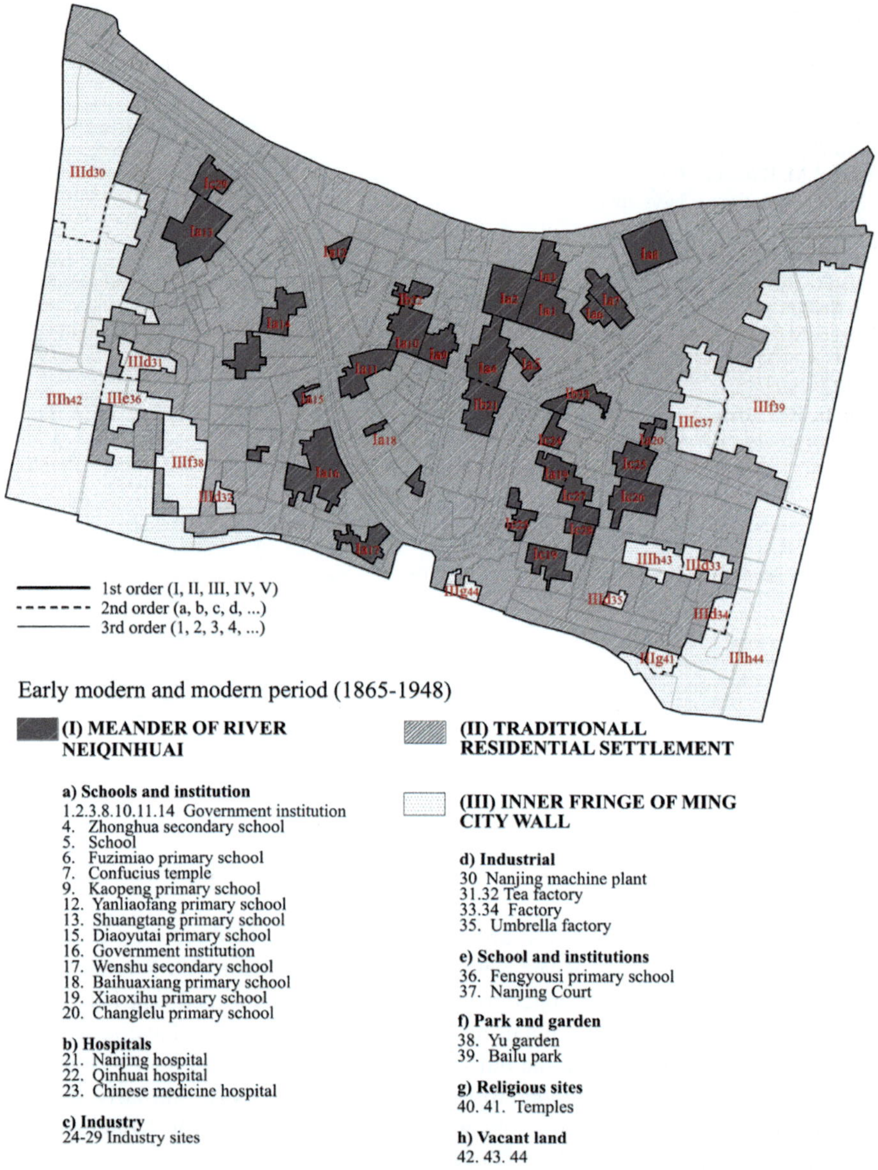

Fig. 6.6 Plan unit division in the early modern and modern period (1865–1949)

Planned economy period (1949–1978)

In this period, the street structure remained largely unchanged except for the widened Changle Road and Jiqing Road for the increasing traffic. Along the widened streets, some of the roadside traditional plots were amalgamated for construction of public buildings (Liu & Wu, 2006).

The industrial production and social organization greatly influenced the urban landscape. Transformations took place mainly on the vacant land along the inner fringe of the Ming city wall (III), where factories were built up on previously vacant land, agriculture land and religious land (Fig. 6.7). This phenomenon is consistent with Conzen's observation of fringe-belt 'absorption' or 'alienation', which is defined as 'the transfer of part of whole of a plan-unit from one integument to another of different character without change of site' (Conzen, 1960, p. 123). Work-units (*danweis*) as a state-owned enterprise providing both working facilities and welfare housing mushroomed in the 1950s (Xie & Costa, 1991). They were located along the outer fringe of the Ming city wall or scattered within the walled city. In Old Chengnan District, the fine-grained traditional residential settlement still remained. The development in this period involved a minimum of destruction in the previously established built-up area within the wall.

Similar with the previous morphological period, the plan unit division is shown in Fig. 6.8.

Transition period (1979–1997)

After the economic reform initiated in 1978, street networks in Old Chengnan District have been greatly updated (Fig. 6.9). The significance of the Ming city wall as a physical barrier was greatly reduced, and the spatial connections of Old Chengnan District with the rest of the city were greatly improved due to these new constructed and straightened streets.

Housing provisions for the growing population were gradually changed from work-units to modern residential units by the government-developer coalition. Some (IVj37,54) were located on the vacant land or open space inside and outside the Ming city wall. Some (IVj36,46,47,50,52) were built up along the major streets, such as Jiqing-Changle Road and Shengzhou-Jiankang Road (Fig. 6.9). Among them, Laifeng residential unit (IVj36) was one of the largest, and it took up 60 ha of previous built-up land by plot amalgamation. Modern apartments and grid streets replaced previous courtyard buildings and small alleyways. The building arrangement was monotonous: slab buildings with the long side oriented north–south. There was similar distance in between each to ensure enough sunlight. Most of the residential units were enclosed by walls and fences.

Tourist-led urban redevelopment (IVk) took place at the Confucius temple area at the end of the 1980s. It was planned and constructed in the traditional building style of the Ming and Qing dynasties. The street alleyways were maintained in original scale. Yet it was no longer a place for study, but rather a hustle-bustle tourist spot.

Plots along Zhonghua Road (IVl) were amalgamated for large-scale commercial and business redevelopment under the policy of 'compensating land for public utility construction' introduced by the Nanjing government; this exchanged rights with real

Fig. 6.7 Plan unit division in the planned economy period (1949–1978) (based on the 1951 cadastral map)

estate developers in land for agreed amounts of services from the Nanjing government. This redevelopment was accompanied by the widening of Zhonghua Road in 1994. These newly built modern building blocks lining the Zhonghua Road were huge compared with the traditional residences. These buildings were arranged with their long side facing the street, and the building depth was around 15–20 m.

Figure 6.10 shows the plan unit division. Apart from previous three first-order plan units, another plan unit—transitional period settlement (IV)—emerged. The second-order boundaries identify different building blocks, such as Mendong traditional settlement (IIe), modern residential units (IVj) and linear commercial development unit (IVl). The third-order boundaries distinguish individual establishments. Yet it is difficult to further subdivide the homogenous traditional residential settlement (IIe and IIf) due to the lack of historic records.

Market economy period (1998–2015)

During this period, the street network structure remained consistent with that in the previous morphological period, although Zhongshan Road was extended southwards in 1997. A new gate (Changan gate) was opened to strengthen the connection between inside and outside of the wall.

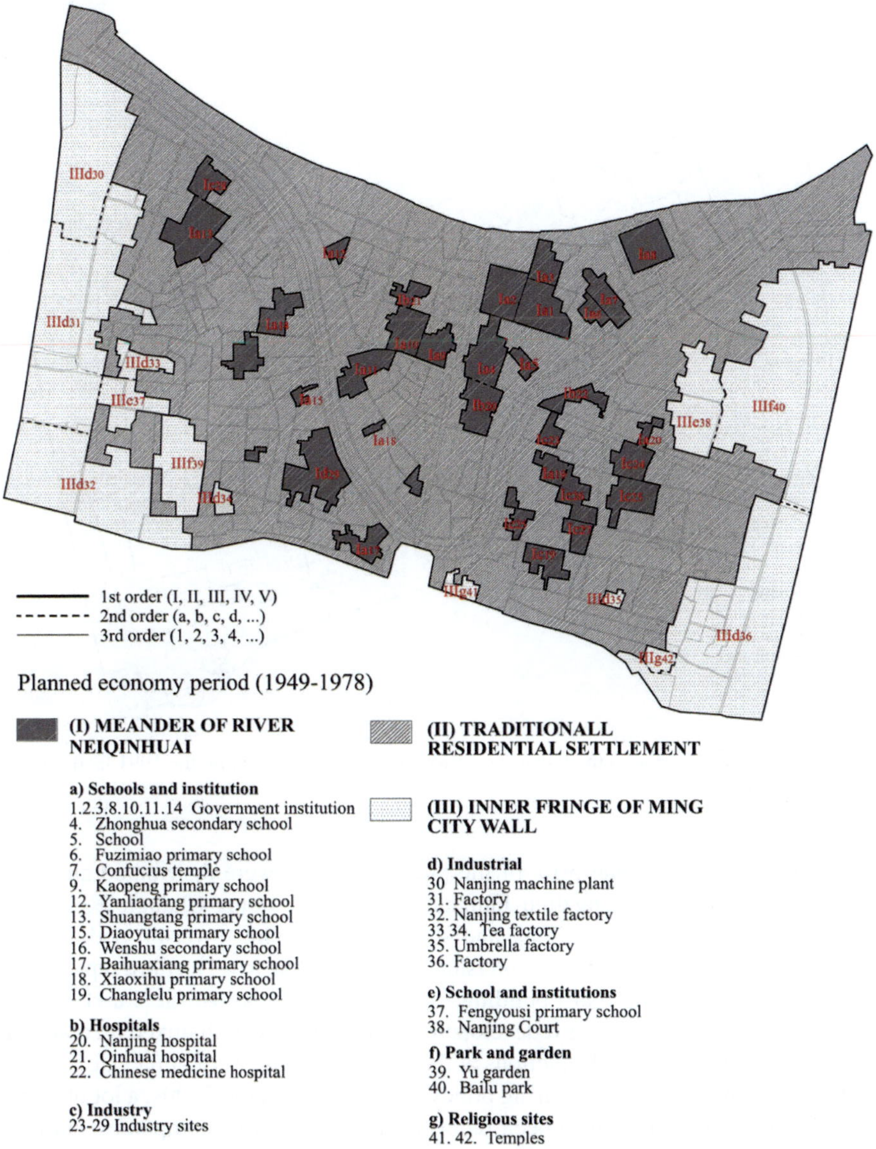

Fig. 6.8 Plan unit division in the planned economy period (1949–1978)

The property-led housing development was located mainly along the River Neiqinhuai and the newly widened Madao Street. Slab building block (Vn61) was the major type of residential building, but the building density was slightly lower than in the previous morphological period (Fig. 6.11). The property-led commercial development manifested in three types. The first type (Vo71) was the large building blocks

Fig. 6.9 Plan unit division in the transition period (1979–1997) (based on the 1991 land use map and 1988 aerial map)

which occupied more than 50% of the plot area. They were distributed along the major roads, such as Shengzhou-Jiangkang Road and Jiqing-Changle Road. The second type (Vp73) was the long and shallow plots, with tower-on-podium type buildings, along the newly built Zhongshan Rd and Xianhe Rd. The third type (Vq75) was the small-scale courtyard buildings in pseudo- or faux-historical architectural style, for example, Yajule was redeveloped into a tourist site with mixed recreational, commercial and residential functions at the east side of the Zhonghua gate in 2007.

Since the industrial restructuring policy was initiated in the 2000s, a lot of manufacturing industrial sites have been redeveloped into creative companies, start-ups or housing estates, such as the Software Industrial Park (IIIg25) developed from the previous Nanjing Machine Tool Plant and the Culture and Creative Industrial Park (IIIg26) from the Nanjing textile industry site. In addition, traditional residential settlement has been further demolished and made way for real estate development in Old Chengnan District since 2000s. Only a few historic areas in Menxi and Mendong remain, such as Hehuatang (IIf22), Diaoyutai (IIf23) and Xiaoxihu (IIe19) remained.

Figure 6.12 shows that the plan unit division. In comparison with previous plan units, one more first-order plan unit—market economy period settlement (V)—emerged. The second-order boundaries highlight different building typologies, such

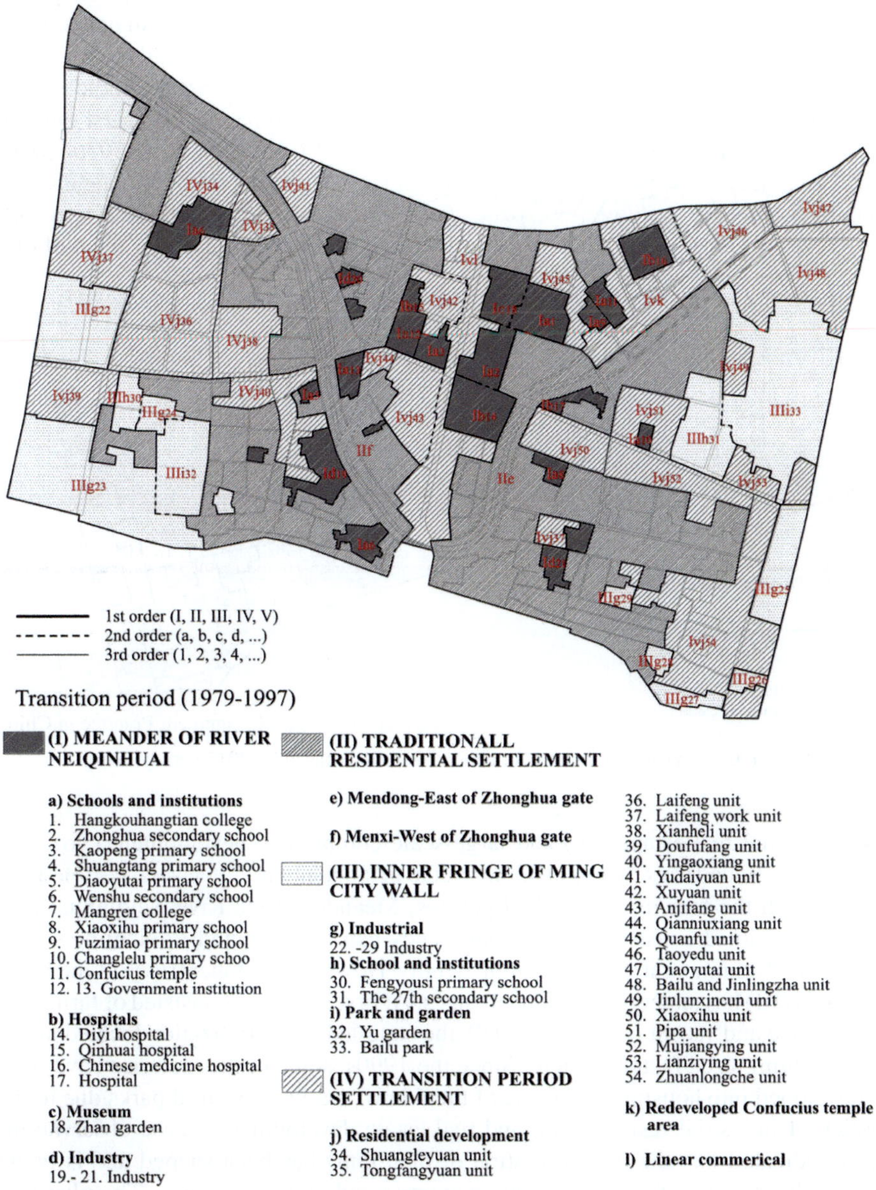

Fig. 6.10 Plan unit division in the transition period (1979–1997)

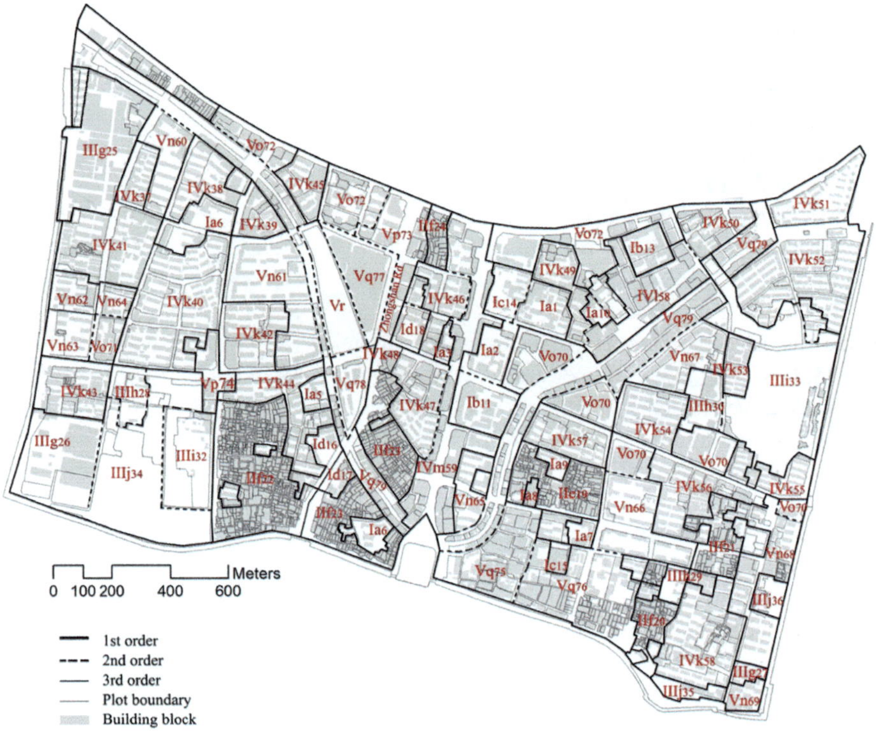

Fig. 6.11 Plan unit division in the market economy period (based on 2015 survey map)

as residential real estate (Vn) and large-scale commercial development (Vo). The third-order boundaries distinguish individual establishments.

Through the four morphological periods, Meander of the Rive Neiqinhuai area (I) has remained consistent. This consistency can be attributed to Nanjing's long established administrative and cultural identity. However, the inner fringe of the Ming city wall (III) has undergone a lot of changes. At first, it mainly consisted of farmland, open space and vacant land. After 1949, the state-led industrialization located large- and medium-sized factories there. Since the 1990s, some of the factories have been redeveloped into housing, and some of them into creative industrial parks due to the policy of industrial restructuring and real estate development. These observations have indicated that the transformation of plan units has been shaped and is being shaped by dynamic forces in social, economic and cultural spheres. The plan unit of traditional residential settlement (II) remained largely unchanged in the first two periods. After the economic reform in 1979, the size of this plan unit kept decreasing, making way for the newly built plan units. Now there are only a few fragmented and isolated patches within the plan unit.

Observing the location of new plan units shows that they were likely to form along the axial streets created from widening and straightening previous street alleyways.

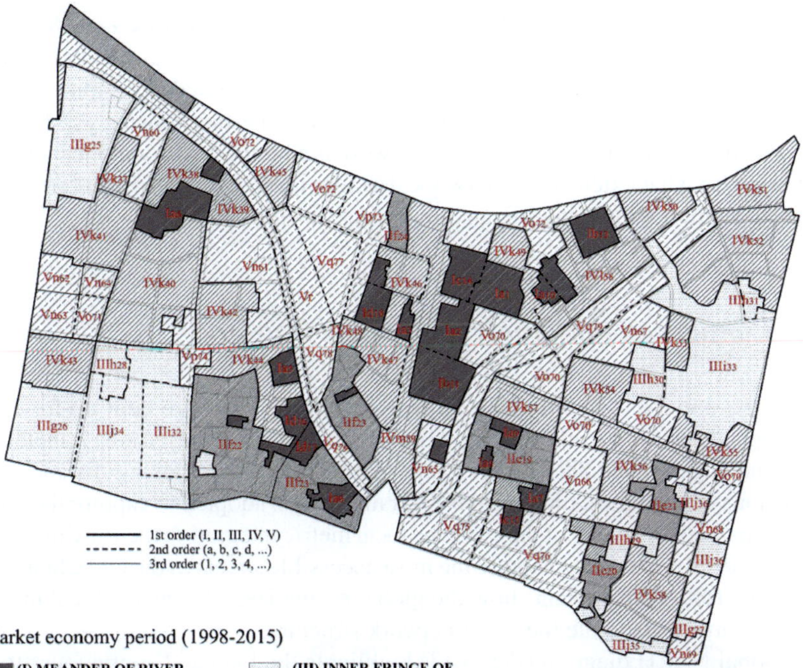

Market economy period (1998-2015)

(I) MEANDER OF RIVER NEIQINHUAI

a) Schools and kindergarten

1. Hangkonghangtian xueyuan 航天航空学院
2. Zhonghua zhongxue 中华中学
3. Kaopeng xiaoxue 考棚小学
4. Shuangtang xiaoxue 双塘小学
5. Diaoyutai xiaoxu 钓鱼台小学
6. Wenshu zhongxue 文枢中学
7. Mangren xuexiao 盲人学校
8. Xiaoxihu youeryuan 小西湖幼儿园
9. Xiaoxihu xiaoxue 小西湖小学
10. Fuzimiao xiaoxue 夫子庙小学

b) Hospitals

11. Diyi yiyuan 市第一医院
12. Qinhuai yiyuan 秦淮医院
13. Zhongyiyuan 中医院

c) Museum

14. Zhanyuan 瞻园
15. Jinlin meishuguan 金陵美术馆

d) Industry

16. Shuguang huagongjituan 化工集团
17. Chacheng 茶城
18. Chinamobile company 电信集团

(II) TRADITIONAL RESIDENTIAL SETTLEMENT

e) Mendong - East of the Zhonghua gate

19. Xiaoxihu unit 小西湖历史街区
20. Santiaoying unit 三条营历史街区
21. Madaojie unit马道街街区

f) Menxi - West of the Zhonghua gate

22. Hehuatang and unit 荷花塘历史街区
23. Diaoyutai unit 钓鱼台历史街区
24. Heilangxiang unit 黑廊巷街区

(III) INNER FRINGE OF MING CITY WALL

g) Redeveloped industrial sites

25. Software industrial park, previous Nanjing Machine Tool Plant 国创园
26. Creative industrial park previous Nanjing cotton mill 悦动新门西
27. Creative industrial park无为创意产业园

h) School and institutions

28. Fengyousi xiaoxue 风游寺小学
29. Foreign language school 外国语中学
30. 27th zhongxue 二十七中学
31. Teacher's school 教师进修学校

i) Park and facilities

32. Yu garden 愚园
33. Bailu park 白鹭公园

j) Vacant land

34. West of Yuyuan 愚园以东
35. South of Renhouli仁厚里小区以南
36. Along Jiangning Rd 江宁路两侧

(IV) TRANSITION PERIOD SETTLEMENT

k) Residential development

37. Lingjiaoshi unit 菱角市小区
38. Shuangleyuan 双乐园小区
39. Tongfangyuan 铜坊苑小区
40. Laifeng unit 来凤小区
41. Laifeng work unit 来凤新村
42. Xianheli 仙鹤里
43. Doufu unit 豆腐坊
44. Yingaoxiang and Jiqing unit 殷高巷集庆小区
45. Yudaiyuan unit 玉带园小区
46. Xuyuan unit许苑小区
47. Anjifang unit 鞍辔坊小区
48. Qianniuxiang unit 牵牛巷小区
49. Quanfu unit 全福小区
50. Taoyedu unit 桃叶渡小区
51. Diaoyuxiang unit 钓鱼台小区
52. Bailu and Jinlingzha unit 白鹭金陵闸小区
53. Jinlun new unit金轮新村
54. Pipa unit 琵琶小区
55. Lianziying unit 莲子营小区
56. Mujiangying unit 木匠营小区
57. Xiaoxihu unit 小西湖小区
58. Zhuanlong che unit 转龙车小区

l) Redeveloped Confucius temple area

m) Linear commercial development

59. Area along Zhonghua Rd 中华路沿线

(V) MARKET ECONOMY PERIOD SETTLEMENT

n) Residential real estate

60. Taoyuan renjia 桃园人家
61. Huangce jiayuan unit 皇册家园
62. Dalai unit 达莱花园
63. 1st Laifeng unit 来凤街一号
64. Jinsu unit 金粟佳园
65. Hongtushang and Huajing unit 宏图上花园
66. Shangfu yuan unit 尚福园
67. Yipin jiayuan unit 一品家园
68. Changle huayuan unit长乐花园 逸景园
69. Qinhuai huayuan unit秦淮花园小区

o) 1990 large-scale commercial development

70. Areas along Changle Rd
71. Fengchao creative industry 凤巢文化创意
72. Areas along Shengzhou and Jiangkang Rd

p) 2000 linear commercial development

73. Area along Zhongshan Rd 中山路沿线
74. Area along Xianhe Rd 仙鹤路沿线

q) 2000 small-scale commercial development

75. Yajule 雅居乐
76. Old Mendong 老门东
77. Areas along the River Neiqinhuai 内秦淮沿线
78. Leduhui shopping cente 中山乐都汇购物中心
79. Jiqing Rd commercial street 集庆路商业街

r) Vacant land

80. Taihe yuanzi 泰和院子

Fig. 6.12 Plan unit division in the market economy period (1998–2015)

For example, in the early 1990s, the residential housing (IVj) was located along the widened Jiqing Road, while the commercial development (IVl) was lined along the widened Zhonghua Road. In the late 1990s, the linear commercial development (Vp) was lined along the South Zhongshan Road while the real estate housing was located along the widened Madao Road. As street network is an important component of plan unit division, to what extent street network structure is related to plan unit formation and transformation will be investigated in the next section.

6.2.2 Configurational Analysis of Street Network

Based on the previous ground plans, street segment models were built up for each morphological period in ArcGIS. A 2 km buffer area from the focused area is included to avoid the edge effect. This 2 km is approximately 25-min walking distance. As with previous case studies, the global NACH measure is adopted to capture the most frequently passed-by streets of the city. The local metric integration measure in 400 m walking distance is chosen to capture the most accessible streets in the Old Chengnan District. The following explains how the global normalized choice and local metric integration further elaborate the street network structure.

The global NACH diagram (Fig. 6.13) highlights the long and continuous street lines which are generally thoroughfares of the city. In the planned economy period, some of the previous street alleyways were straightened and widened into east–west Changle Road, the north–south Zhonghua Road. Since the transition period, the thoroughfares of the city have grown and extended, such as the southern extension of Zhongshan Road, the widened Madao Road and Gutong Road (Fig. 6.14). The transformation of the street network can be viewed a means to accelerate movement to overcome the increasing distance between the centre and the edge.

In the local integration diagram, the red patchworks are within the traditional residential settlement. Figure 6.15 shows that these areas have smaller building blocks

Fig. 6.13 Global measure of the street network of Old Chengnan District in early modern and modern period, planned economy period

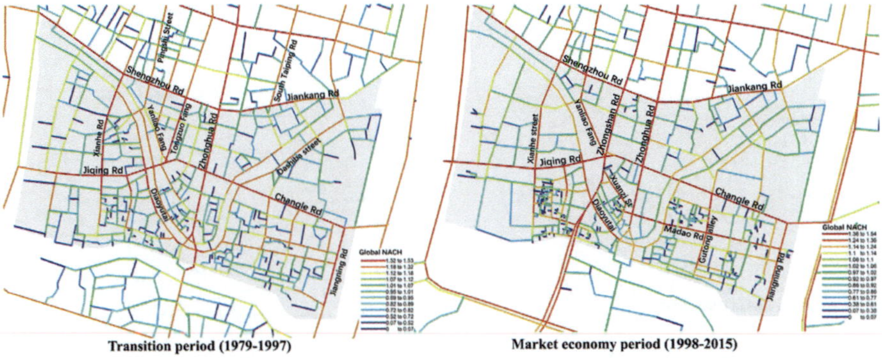

Fig. 6.14 Global measure of the street network of Old Chengnan District in transition period and market economy period

and shorter distance from one street segment to another. They visually corresponded to the clustered traditional residential areas in the Ming and Qing dynasties, such as Lingjiaoshi, Diaoyutai, Hehuatang and Yanliaofang. These areas were characterized by a higher density of alleyways, smaller plots and courtyard buildings, which remained rather consistent in the first two periods. Since the economic reform, the scale of identifiable patchworks has shrunk greatly, and some were demolished due to new road construction (Fig. 6.16). The few patchworks correspond to the named historic areas in the 2012 Guidelines on Conservation Planning of Precious Historic and Cultural Cities of Nanjing, such as Hehuatang, Diaoyutai, Santiaoying, Confucius temple and Xiaoxihu. It could be tentatively argued that the configurational approach is able to identify subareas with distinctive historical and cultural characteristics, which were shaped in the Feudal period by dynamic social, economic and cultural forces.

The configurational approach could reveal the generative rules which contribute to the growth of cities: the global NACH uncovered the skeleton of the city, which

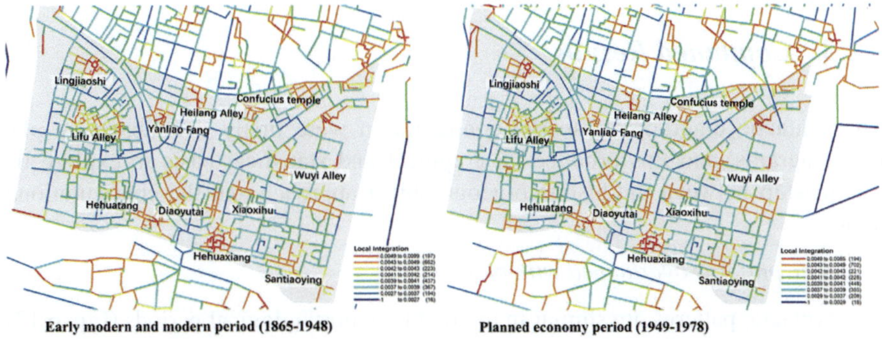

Fig. 6.15 Local measure of the street network of Old Chengnan District in early modern and modern period, planned economy period

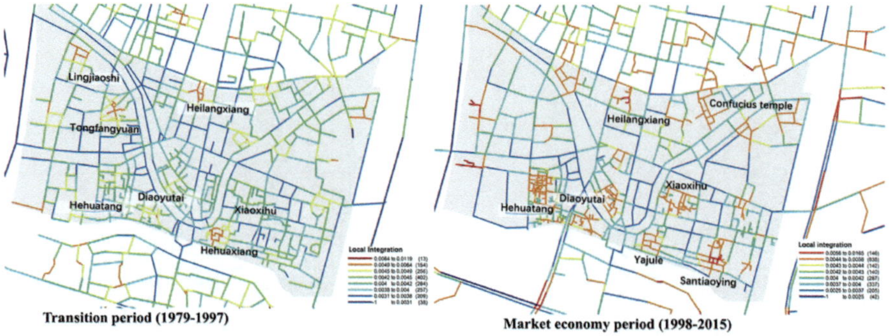

Fig. 6.16 Local measure of the street network of Old Chengnan District in transition period and market economy period

overcomes locality and links the patchwork identified by local metric integration into a whole; the local metric integration identified subareas with distinctive historical and cultural characteristics. This 'dual structure' made up of foreground and background street network could exemplify the generative rules of space in Nanjing, that is cities comprise a small number of longer lines with near straight connections that seek to maximize movement driven by microeconomic factors, and a large number of shorter lines with near right-angle connections that are configured to restrain and structure movement driven by social and cultural factors (Hillier, 1996).

Meanwhile, the results of both the global and local measures are then investigated statistically. As with previous case studies, the street segments with the top 20% of the global NACH values and the upper fifty percent of the local integration values could better describe the integration core and local cluster of street segments of Nanjing across the four morphological periods.

The next section will combine the analytical results of the historico-geographical and configurational approaches and examine the extent to which the spatial configuration contributes to plan unit recognition.

6.2.3 Combined Results

The syntactic diagrams of spatial configuration were firstly overlapped on plan unit diagrams, and this produced a graphical and statistical basis to explore the relationship between the historico-geographical approach and the configurational approach.

Syntactic patterns overlapped with plan units

The syntactic patterns are similar in the first two morphological periods (Fig. 6.17). The red lines show that the first-order plan units of Meander of the River Neiqinhuai (I) and Traditional settlement (II) are located along the most integrated streets. In the transition period and market economy period, the newly emerged first-order plan units (IV) are organized along the integration core (Fig. 6.18). One possible explanation is the redevelopment of a settlement tends to locate along the major roads of the city to take advantage of movement flows.

A number of the local clusters comprised of the orange lines correspond to the third-order plan units (Figs. 6.17 and 6.18), such as Lingjiaoshi, Diaoyutai, Hehu-atang and Yanliaofang in the traditional residential settlement (IIe and IIf). In the market economy period, apart from previous residential settlement, the local cluster of the street segments could capture some emerging new patchworks in the post-2000 small-scale commercial development (Vq). These highly accessible streets are more likely to provide good accessibility to sustain and support the functioning of these urban areas.

The results of the pattern comparison suggest that the street configuration measured by the global NACH is correlated with the distribution of the first-order plan units across the four morphological periods. Yet the street configuration measured by the local integration may not capture all the third-order plan units. Other factors like development policies and strategies may outweigh the impact of street configuration in the formation of plan units. The compositions of plots and building blocks may play a stronger role in characterizing urban settlement than street networks alone. This is similar with the observation in Chinatown, Singapore.

Syntactic values of plan units

The aggregate values of both global NACH and local integration are measured for plan units in each morphological period. Their values are summarized and related to the historical urban process of Old Chengnan District. In real cities, the mean NACH value varies from about 0.6 to 1.3. The increasing NACH values indicate that this whole district has become more integrated into the city (Table 6.1). This can be

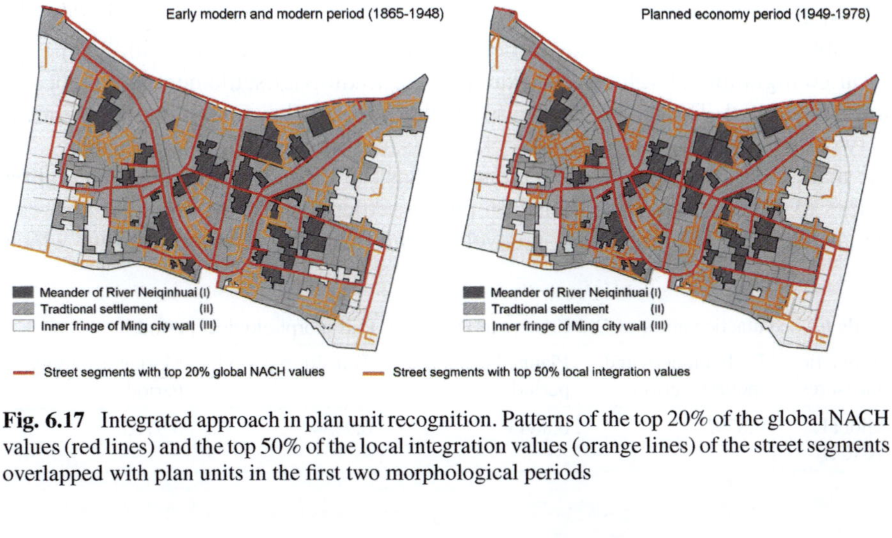

Fig. 6.17 Integrated approach in plan unit recognition. Patterns of the top 20% of the global NACH values (red lines) and the top 50% of the local integration values (orange lines) of the street segments overlapped with plan units in the first two morphological periods

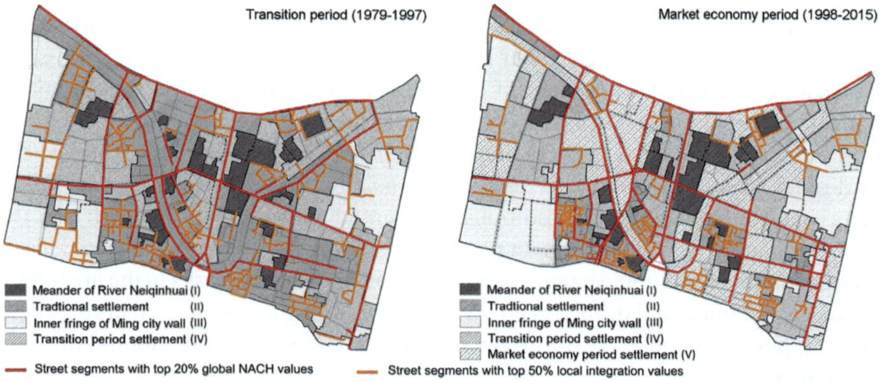

Fig. 6.18 Integrated approach in plan unit recognition. Patterns of the top 20% of the global NACH values (red lines) and the top 50% of the local integration values (orange lines) of the street segments overlapped with plan units in the last two morphological periods

attributed to the rapid urban growth outside the Ming city wall since the economic reform, making this district well embedded in the city.

In Table 6.2, there are two observations: firstly, the newly developed plan units tend to have higher global NACH values, for example, the transition period settlement (IV) and market economy period settlement (V). That is because the urban redevelopment in Old Chengnan District mainly replace previous urban fabric, which became much more embedded as the city grew. This situation is similar with that observed in Chinatown, Singapore. Secondly, most of the old plan units may have increased global NACH values as the city grows despite one plan unit—the traditional settlement (II). This could be explained by the fact that areas along the major streets in this plan unit were replaced by other emerging new plan units, it became spatially segregated from its surroundings, which may result in the loss of its spatial advantage within the old city.

Table 6.3 shows that the lower-tier plan units are characterized with distinctive local configurational values. For example, the traditional settlement (II) is locally most integrated. The underlying reason is that this area has a high-density of street network and small-scale building blocks and is most accessible to local residents. Yet the market economy period settlement (V) has the second lowest local integration values (4.13E−3), so is the transition period settlement (IV) (4.16E−3). This can be attributed to the large-scale redevelopment since the economy reform, which

Table 6.1 Syntactic values of Old Chengnan District in four morphological periods

Syntactic measures	Early modern and modern period	Planned economy period	Transition period	Market economy period
Global NACH	0.8792	0.8843	0.9334	0.9105

Table 6.2 Global NACH values of 1st tier plan units in four morphological periods

Global NACH of plan units	Modern period 1865–1948	Planned economy period 1949–1978	Transition period 1979–1997	Market economy period 1998–2015
I. Meander of River Neiqinhuai	0.8917	0.9050	1.0303	1.0632
II. Traditional settlement	0.8872	0.8880	0.9280	0.8221
III. Inner fringe of Ming city wall	0.8102	0.8567	0.8644	1.0038
IV. Transition period settlement			1.0200	1.0019
V. Market economy period settlement				1.0528

is characterized by the modernist city planning of large block size and low street network density.

The value comparisons show that the historico-geographical approach could provide a place-specific context for interpreting the configurational values of plan units and examine the extent to which the generative rules of spatial configuration are related to the formation and transformation of plan units.

From this case study, it can also be seen that combining the historico-geographical and configurational approaches provides a richer understanding of the historical transformations of urban form: it is conditioned by the external socio-economic factors; at the same time, it is also underscored by the more generative rules of urban space.

Apart from the historic character of built environment, researchers have argued that the historic district could serve as a powerful catalyst for socio-economic regeneration (Bandarin & Van Oers, 2012; Jacobs, 1961; Koh & Melic, 2017). The development of diverse economic activities and an improvement in the quality of life could be achieved through sensitive small-scale, incremental modifications of form and function as a source of changes (Fang et al., 2012; Scheer, 2010; Shan, 2006; Wang & Mo, 2017; Wu, 1994; Zhang et al., 2011). Therefore, understanding how the existing socio-economic activities are organized in the built environment and what the potentials are is important for future interventions. The next section explores

Table 6.3 Local metric integration values of 1st tier plan units in four morphological periods

Local metric integration of plan units	Modern period 1865–1948	Planned economy period 1949–1978	Transition period 1979–1997	Market economy period 1998–2015
I. Meander of River Neiqinhuai	4.00E−03	4.00E−03	3.97E−03	4.08E−03
II. Traditional settlement	4.19E−03	4.19E−03	4.24E−03	4.59E−03
III. Inner fringe of Ming city wall	4.04E−03	4.07E−03	4.20E−03	4.17E−03
IV. Transition period settlement			4.16E−03	4.32E−03
V. Market economy period settlement				4.13E−03

combining the historico-geographical approach and the configurational approach to evaluate the character and strength of landscape units in terms of historicity and spatial potential.

6.3 Landscape Unit Evaluation

This section mainly focuses on contemporary functional values and their implications of the future. Thus, landscape units of the current period—market economy period (1998–2015)—are studied by integrating the mappings of building fabric and land use together with previous plan unit division in Sect. 6.3.1. The configurational approach was applied in Sect. 6.3.2 to investigate the spatial potentials of landscape units, that is to what extent urban form could support, impede or organize socio-economic activities. Section 6.3.3 combines the results of both approaches and evaluates both the character and strength of landscape units, which is valuable information for future development.

6.3.1 Historico-geographical Analysis

As is discussed in Sect. 2.1.3, the plan unit division of the market economy period (Figs. 6.11 and 6.12) contains a three-tier hierarchy of units ranging from major divisions of the town to very localized ones. It is highly resistant to change and forms 'the framework for the other man-made features', which are building block and land use (Conzen, 1973).

The building fabric of Chengnan District shows distinct juxtapositions and groupings of buildings from different periods (Fig. 6.19). Traditional residential buildings created in the Ming and Qing dynasties still exist today. They are one- or two-storey high in courtyard layout. Residential buildings built between 1979 and 1990 are characterized by multi-storey slab blocks. Residential buildings built after the 1990s have tended to be denser, either in high-rise slab or tower blocks. Commercial buildings constructed in the late 1980s are mainly in pseudo-traditional style, driven by tourism or culture-led revitalization. Since the 1990s the real estate commercial development has contributed to a variety of building forms: large commercial building blocks, commercial streets with groups of pseudo-traditional building blocks, as well as tower-on-podium form for mixed uses, such as residence/commerce and office/commerce. Institutional buildings are loosely arranged within the plot. Consistent with Conzen's (1973) argument, building fabric could be easily adapted, rebuilt or replaced by new buildings, which are less persistent to change than plan unit. It mainly partakes in the 2nd and 3rd tier landscape units.

Figure 6.20 shows that the major type of land utilization in Old Chengnan District is residential use, including the traditional settlement and modern apartments built after the 1980s. Commercial use is mainly concentrated along Zhonghua Road, and

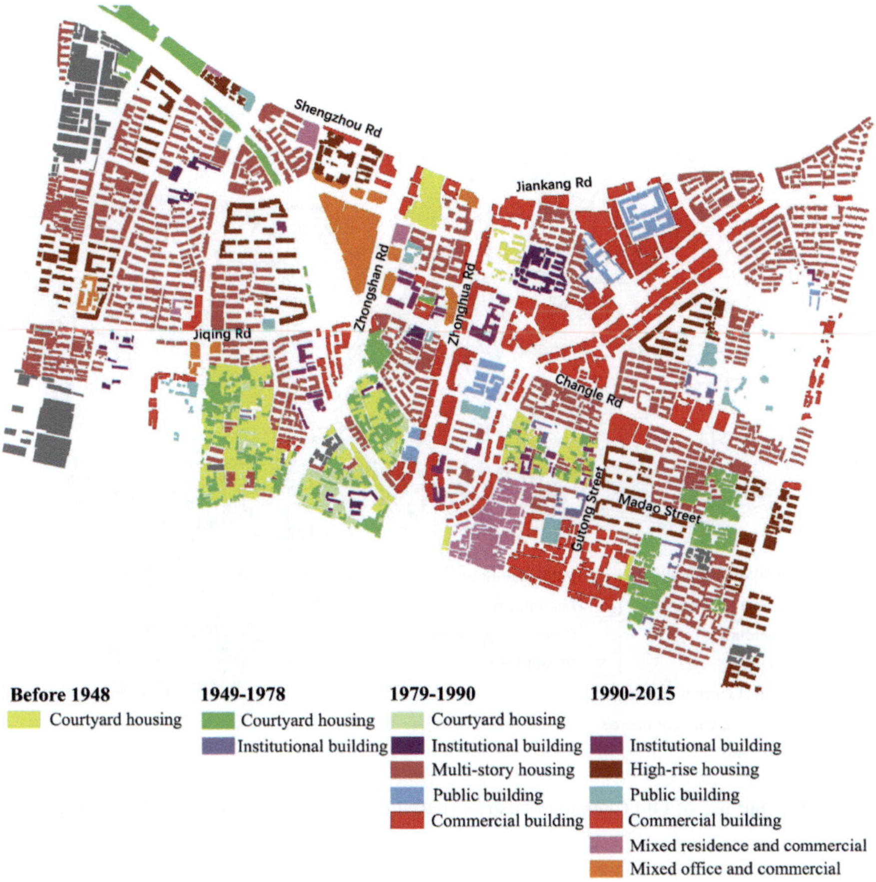

Fig. 6.19 Building type of Old Chengnan District (based on site survey, Google Map and survey map of Nanjing in 2015)

Qinhuai River and in the Confucius temple area. Industrial land use is mainly located at the fringe of the Ming city wall and some are scattered in the residential area. Institutions and public facilities are mainly settled along the River Neiqinhuai. There are a few mixed uses, such as mixed residence with commerce and mixed office with commerce. They are located along Zhonghua Road and Zhongshan Road. Compared to plan unit and building fabric, land use is more ephemeral to change and its contribution to landscape units lies in the lowest rank, which is the 4th tier in this study.

The three maps together are the basis of a fourth, composite map of urban landscape units, also comprising a four-tier hierarchy of units (Fig. 6.21). The map of urban landscape units provides the basic framework for analysing how individual forms relate to their surroundings and articulating conservation proposals. Rather than focusing on individual buildings that are worthy of preservation, the future of

Fig. 6.20 Land use of Old Chengnan District (based on detailed land use planning of Nanjing in 2012)

present forms should be assessed in relation to the character that urban landscape units represent.

The historico-geographical approach attaches great importance to the historicity of urban landscape. Land use helps to understand how to present socio-economic activities are distributed spatially. It has many categories and is classified into mixed, bi and mono, indicating the different degrees of socio-economic activities. Yet it is at a coarse level of resolution, which fails to include the emerging multiple functions within buildings, either vertically or horizontally as city evolves. Take IIf for example; for the Traditional residence (R3), the mono-land use suggests homogeneous socio-economic activities in this landscape unit. However, in fact, some buildings along the alley ways in Hehuatang unit have different amenities and services on the ground floor.

There is also a problem with the land use unit boundaries. According to Conzen, plot is defined as 'a parcel of land representing a land use unit defined by boundaries on the ground' (Conzen, 1960, p. 128). Plot is both the physical form and the land use unit. In Fig. 6.22 (second from left), the land use map of Old Chengnan District in

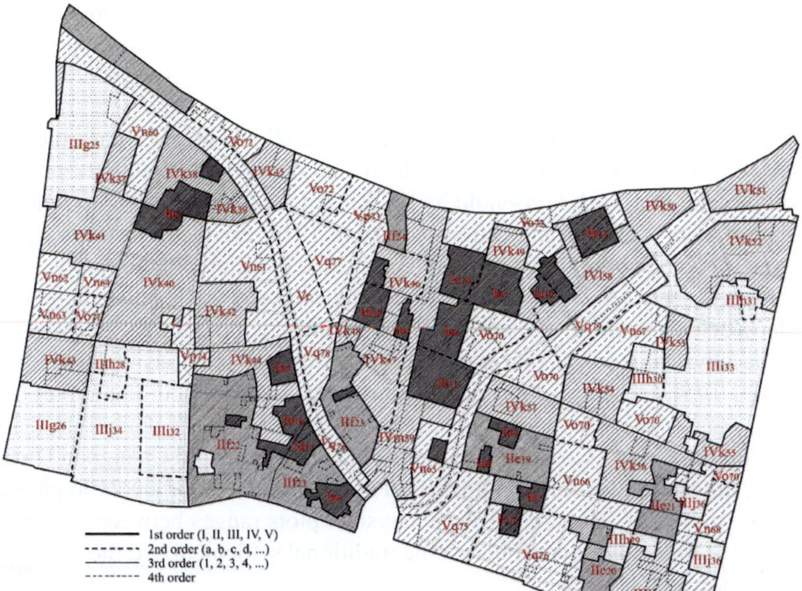

Market economy period (1998-2015)

(I) MEANDER OF RIVER NEIQINHUAI

a) Schools and kindergarten

1. Hangkonghangtian xueyuan 航天航空学院
2. Zhonghua zhongxue 中华中学
3. Kaopeng xiaoxue 考棚小学
4. Shuangtang xiaoxue 双塘小学
5. Diaoyutai xiaoxu 钓鱼台小学
6. Wenshu zhongxue 文枢中学
7. Mangren xuexiao 盲人学校
8. Xiaoxihu youeryuan 小西湖幼儿园
9. Xiaoxihu xiaoxue 小西湖小学
10. Fuzimiao xiaoxue 夫子庙小学

b) Hospitals

11. Diyi yiyuan 市第一医院
12. Qinhuai yiyuan 秦淮医院
13. Zhongyiyuan 中医院

c) Museum

14. Zhanyuan 瞻园
15. Jinlin meishuguan 金陵美术馆

d) Industry

16. Shuguang huagongjituan 化工集团
17. Chacheng 茶城
18. Chinamobile company 电信集团

(II) TRADITIONAL RESIDENTIAL SETTLEMENT

e) Mendong - East of the Zhonghua gate

19. Xiaoxihu unit 小西湖历史街区
20. Santiaoying unit 三条营历史街区
21. Madaojie unit 马道街街区

f) Menxi - West of the Zhonghua gate

22. Hehuatang and unit 荷花塘历史街区
23. Diaoyutai unit 钓鱼台历史街区
24. Heilangxiang unit 黑廊巷街区

(III) INNER FRINGE OF MING CITY WALL

g) Redeveloped industrial sites

25. Software industrial park, previous Nanjing Machine Tool Plant 国创园
26. Creative industrial park previous Nanjing cotton mill 悦动新门西
27. Creative industrial park 无为创意产业园

h) School and institutions

28. Fengyousi xiaoxue 凤游寺小学
29. Foreign language school 外国语中学
30. 27th zhongxue 二十七中学
31. Teacher's school 教师进修学校

i) Park and facilities

32. Yu garden 愚园
33. Bailu park 白鹭公园

j) Vacant land

34. West of Yuyuan 愚园以东
35. South of Renhouli 仁厚里小区以南
36. Along Jiangning Rd 江宁路两侧

(IV) TRANSITION PERIOD SETTLEMENT

k) Residential development

37. Lingjiaoshi unit 菱角市小区
38. Shuangleyuan 双乐园小区
39. Tongfangyuan 铜坊苑小区
40. Laifeng unit 来凤小区
41. Laifeng work unit 来凤新村
42. Xianheli 仙鹤里
43. Doufu unit 豆腐坊
44. Yingaoxiang and Jiqing unit 殷高巷集庆小区
45. Yudaiyuan unit 玉带园小区
46. Xuyuan unit 许苑小区
47. Anjifang unit 鞍辔坊小区
48. Qianniuxiang unit 牵牛巷小区
49. Quanfu unit 全福小区
50. Taoyedu unit 桃叶渡小区
51. Diaoyuxiang unit 钓鱼台小区
52. Bailu and Jinlingzha unit 白鹭金陵闸小区
53. Jinlun new unit 金轮新村
54. Pipa unit 琵琶小区
55. Lianziying unit 莲子营小区
56. Mujiangying unit 木匠营小区
57. Xiaoxihu unit 小西湖小区
58. Zhuanlong che unit 转龙车小区

l) Redeveloped Confucius temple area

m) Linear commercial development

59. Area along Zhonghua Rd 中华路沿线

(V) MARKET ECONOMY PERIOD SETTLEMENT

n) Residential real estate

60. Taoyuan renjia 桃园人家
61. Huangce jiayuan unit 皇册家园
62. Dalai unit 达莱花园
63. 1st Laifeng unit 来凤街一号
64. Jinsu unit 金粟佳园
65. Hongtushang and Huajing unit 宏图上花园
66. Shangfu yuan unit 尚福园
67. Yipin jiayuan unit 一品家园
68. Changle huayuan unit 长乐花园 逸景园
69. Qinhuai huayuan unit 秦淮花园小区

o) 1990 large-scale commercial development

70. Areas along Changle Rd
71. Fengchao creative industry 凤巢文化创意
72. Areas along Shengzhou and Jiangkang Rd

p) 2000 linear commercial development

73. Area along Zhongshan Rd 中山路沿线
74. Area along Xianhe Rd 仙鹤路沿线

q) 2000 small-scale commercial development

75. Yajule 雅居乐
76. Old Mendong 老门东
77. Areas along the River Neiqinhuai 内秦淮沿线
78. Leduhui shopping cente 中山乐都汇购物中心
79. Jiqing Rd commercial street 集庆路商业街

r) Vacant land

80. Taihe yuanzi 泰和院子

Fig. 6.21 Landscape unit of Old Chengnan District in the market economy period (1998–2015)

2012 shows the distinctive land use units, such as R_2-modern housing (in yellow) or M_1-industry (in brown). However, in traditional urban settlement R_3, the land use unit boundary does not correspond to the physical form which consists of predominantly small-scale courtyard buildings.

As shown in the 1951 cadastral map for example (Fig. 5.40 far right), there are subdivisions within this land use unit which tended to reflect the pattern of property ownership. In the plot, each household had control of access and activities, but since the initiation of the housing allocation system in 1949, private land was acquired by the government and buildings on it were allocated to different households. Some households still share one courtyard with the same entrance, while others build walls to divide previous building groups into several units with different accesses. To retrieve the physical form of plot, reference was made to previous cadastral maps of 1937 and 1951 and current physical features, such as walls, in the survey map of Nanjing (Fig. 6.23). This identifiable physical form of plot is defined by Kropf (1997) as 'physical plot'. Compared with the land use plot in Old Chengnan District in Nanjing, the physical plot is at a finer level of resolution. There are 2106 physical plots in the study area. The area of these physical plots ranges between 50 m^2 and 11 ha. Some of the smallest are found in the traditional settlement and the largest are in industrial areas.

Based on the above discussions, a more accurate description of functional use at the scale of physical plot is needed to identify the variations of socio-economic performance at a finer scale. The functional mix model (Hoek, 2010) was combined together with the land use and emerging open urban data POI to explore whether this combination could help to measure the finer-scale socio-economic performance of urban form.

As introduced in Sect. 4.2.2, POI is a set of point data carrying the spatial locations and properties of individual attractions at the building scale (Liu and Long, 2015). It prioritizes small scale and could identify the multiple functions within buildings, either horizontally or vertically with the help of Google Street map. Collected from the online map, data are more up-to-date compared to the 2012 land use data.

According to Hoek's framework discussed in Sect. 4.2.2, urban functions are classified into live, work and visit. Floor area of the three functions is an indicator of the

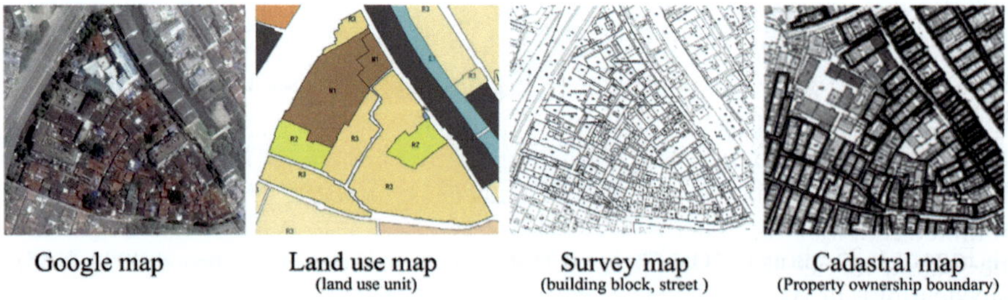

Fig. 6.22 Illustrations of land use unit (2012), building block (2015) and private property boundary (1951)

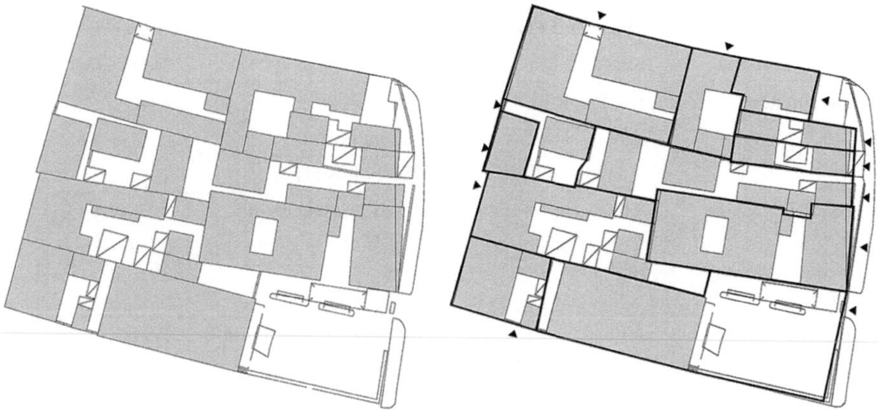

Fig. 6.23 Drawing of physical plot boundary. Building groups with same access were identified, and boundaries of physical plot were drawn

level of socio-economic activities at the building scale. This was calculated using a cadastral map, which contains the building block and floor information, and POIs which carry the functional type. Building floor areas of live/work/visit were accumulated within each physical plot. For calculating the floors areas of live/work/visit functions, previous urban land use classification published in 2012 and POI classification were reclassified into three categories: live, work and visit (Table 6.4).

Secondly, by overlapping POIs with cadastral and land use maps of Nanjing in ArcGIS, three types of combination were generated, (1) land use (live) and POI (live/visit); (2) land use (work) and POI (live/work/visit); (3) land use (visit) and POI (live/visit). In each type, floor areas of live/work/visit functions were calculated separately (Table 6.5).

Type 1 is mainly composed of residential units, hotels, and mixed residence with commerce. Most of these buildings have visit POIs on the ground floor, which usually takes advantage of the apartment rooms facing the street, while the rest of an apartment is still used as residence. By empirical investigations of the different residential typologies shown in Fig. 6.24, the floor area of visit POIs was roughly calculated as 40% of the ground floor area.

Type 2 is mainly composed of institutions, government, schools and mixed office with commerce. Some of the buildings have visit POIs on the ground floor, and live POIs above.

Type 3 mainly consists of supermarkets, libraries, museums, hospitals, financial institutions and religious facilities. Some of the buildings have live POIs above the ground floor.

Figure 6.25 shows the visualization of live/work/visit functions at the scale of physical plot in ArcGIS: the darker the colour, the more building floor areas each physical plot has. The live and visit functions are scattered all over the site while the work function is largely along Zhongshan Road and Zhonghua Road.

Table 6.4 Reclassified land use and POI typology following Hoek's (2010) framework

Land use division	Sub-division			New category
(R) Residence	R1, R2, R3		Residence	Live
	Ra	Raa	Student hostels	
		Rab	Elderly hostels	
	Rb		Mixed residence with commerce	
(B) Commerce and service	B14		Hotel	
(A) Public facility	A1		Government institution	Work
	A3	A31	University and military college	
		A32	Polytechnic school	
		A33a	Primary school	
		A33b	Secondary school	
		A33c	High school	
		A34	Special education school	
		A35	Research institute	
(B) Commerce and service	Bb		Mixed office with commerce	
(M) Industry	M1, M2, M3		Industrial sites	
(R) Residence	Rc		Community center	Visit
(A) Public facility	A2	A21	Library, museum, gallery, exhibition hall	
		A22	Youth and elderly activity center	
	A4		Indoor gymnasium	
	A5		Hospital	
	A7		Heritage site	
	A9		Religious site	
(B) Commerce and service	B1	B11	Retail, mall, market	
		B13	Catering	
	B2	B21	Insurance, bank, stock	
		B22	Television production, media	
	B29a		Research institute	
	B29b		Consultant	
	B3	B31	Theater, music hall, dance hall	
	B4		Postal office, telecom service	
	B9		Vehicle repair shop	

(continued)

Table 6.4 (continued)

POI category	Sub-category	New category
Residence	Residential units	Live
Hotel	Hotels	
Government agency	Government agency	Work
School and institution	School	
	Education organization	
	Research institute	
Business office	Business office	
Cultural facilities	Museum, library	Visit
Leisure and sports	Sports hall	
	Ktv, Majiang, etc.	
Medical service	Pharmacy	
	Hospital	
Catering	Restaurant	
	Coffee shops, teahouse	
Financial services	Atm	
	Bank	
Amenities	Law firms	
	Bath center	
	Hair and beauty salon	
	Customer service center	
	Laundry	
	Travel agency	
	Real estate agency	
Shopping	Vegetable market	
	Supermarket	
	Retails	
	Stores of household appliances	
	Shopping mall	
Service enterprise	Service enterprise	

Hoek (2010) argues that the proportion of live, work and visit functions determines functional mix typology: mono-use, bi-function and mixed. This typology indicates different degrees of the liveliness and movement of the area during day and night. The tertiary diagram (Fig. 6.26) shows that the majority of physical plots have mono-use. There are several bi-functional types, which are live/work and live/visit. Only a few physical plots are mixed live/work/visit.

Figure 6.27 shows the spatial mapping of the functional mix typology of each physical plot in ArcGIS. The rich mix (plots in purple, grey, yellow) lines the major

Table 6.5 Calculations of floor areas of live/work/visit in three land use categories

	Type 1	Type 2	Type 3
Floor areas of live	Total floor area – floor areas of visit	Ground floor area * level (observed from Google Street View)	Ground floor area * level (observed from Google Street View)
Floor areas of work	0	Total floor area – floor areas of visit – floor areas of live	0
Floor areas of visit	40% * building ground floor area	Ground floor area	Total floor area – floor areas of live

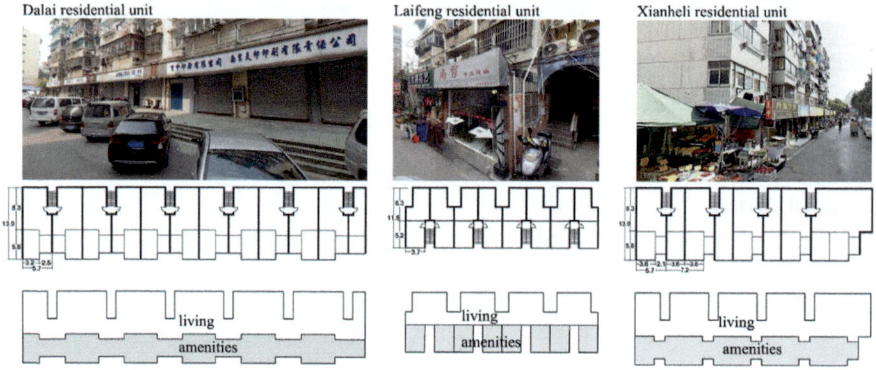

Fig. 6.24 Examples of ground floor function division

north–south Zhonghua and Zhongshan Roads. The mono-visit function (plots in green) is largely concentrated near the Confucius temple area. The mono-live function (plots in red) is scattered all over the site and the mono-work function (plots in blue) is sparsely distributed. Moreover, some subareas are delimited. For example, some physical plots in traditional residential areas (IIe and IIf) are highlighted with live/visit functions. Some in linear commercial development (Vm) are identified with live/work and mixed functions. This shows that the land use and POI data together with Hoek's framework are able to shed light on the understanding of functional diversity through compositions of live/work/visit functions, and effectively map the different functional mix typologies at the scale of physical plot.

6.3.2 Configurational Analysis

This section starts by giving a more accurate and finer-scale description of socio-economic activities. Apart from land use, emerging open data POI was included

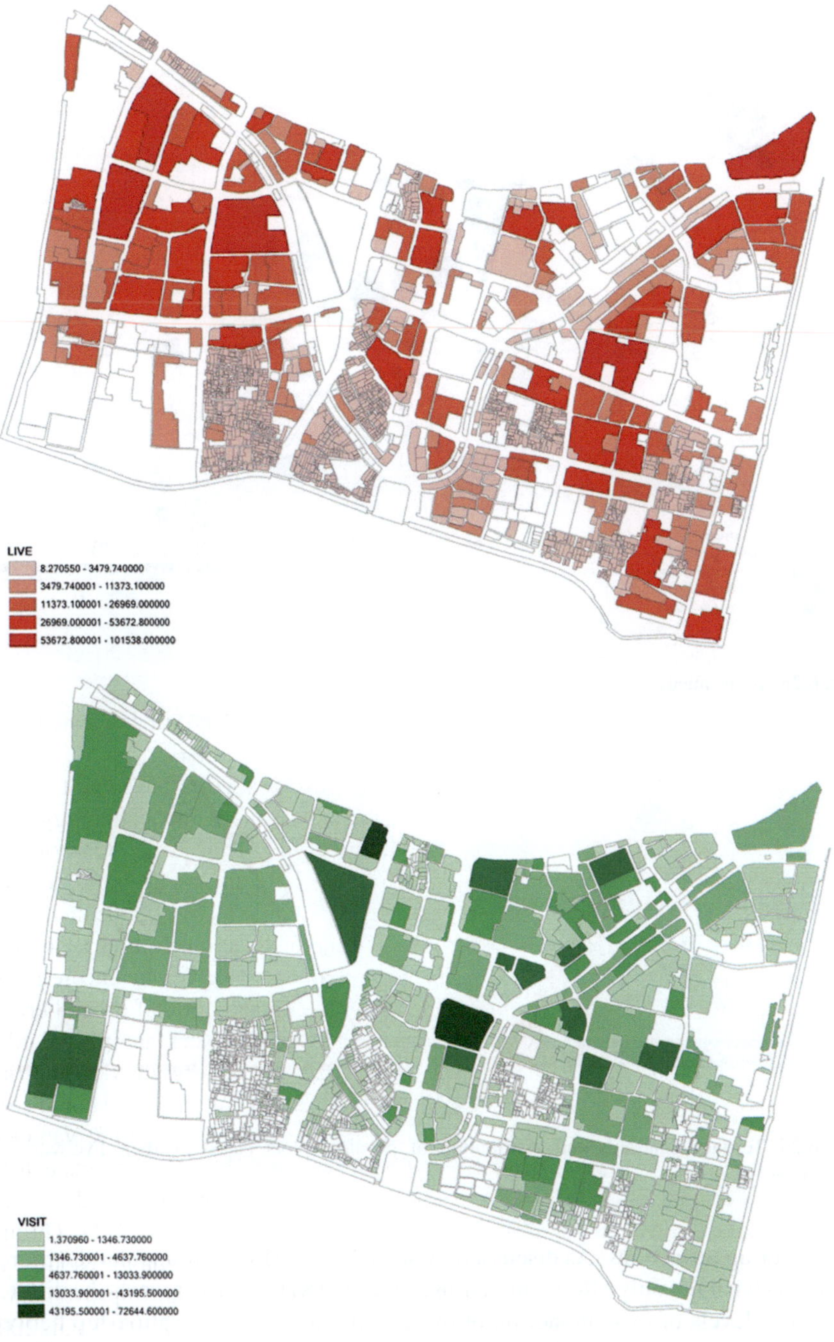

Fig. 6.25 Measurement of the floor areas of live/work/visit functions

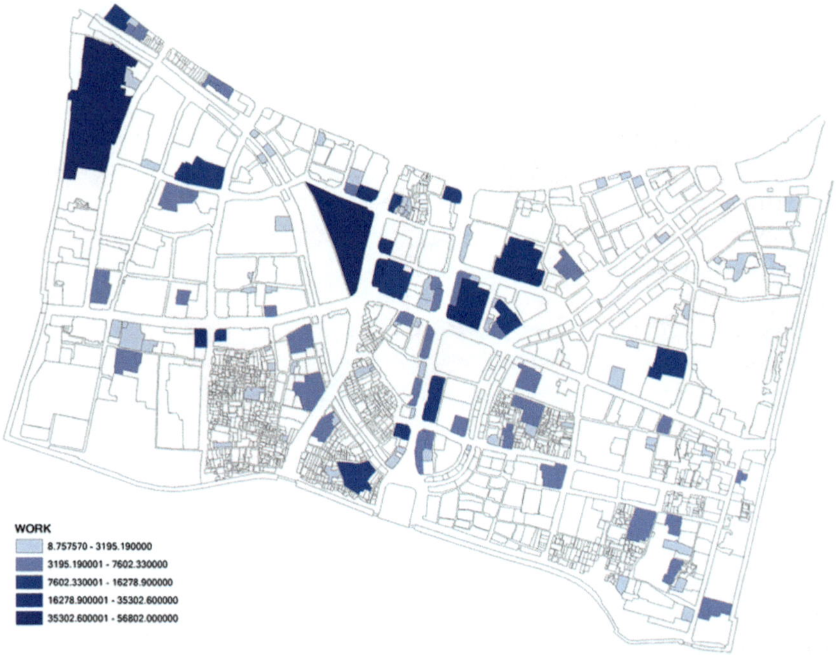

Fig. 6.25 (continued)

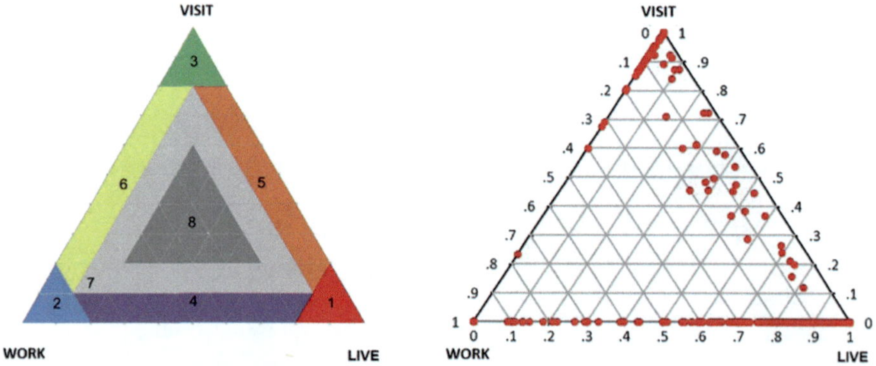

Fig. 6.26 Tertiary diagrams and distribution of physical plots based on their live/work/visit percentage

in measuring attractions. As discussed in Sect. 3.3.2.2,Hoek's triangular diagram of live/work/visit not only offers a mixed-use index (MXI), but also reveals the different kinds and levels of mix (mono, bi, or mixed). It uncovers the relationship between functional pattern and spatial pattern.

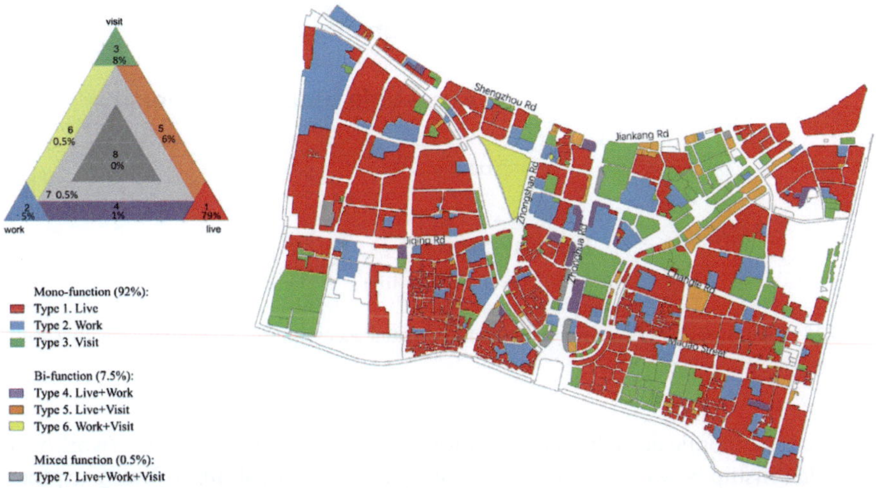

Fig. 6.27 Mapping of functional mix typology based on the floor areas of live/work/visit functions

Apart from interpreting the growth of cities by the generative rules of space, the configurational approach could also uncover the relationship between spatial configuration, urban movement and land use. Hoek (2010) argues, through empirical studies of Dutch cities, that there is a relationship between functional characteristics and physical features, such as accessibility. The more centrally located urban areas are, the more balanced the mix of housing, working and amenities are. Similarly, Hillier (1996) argues that spatial configuration has an impact on socio-economic activities, and that 80% of retails are located along 20% the most spatially accessible streets. These findings suggest that functional diversity and accessibility are related to each other and this relationship might suggest the locational potentials of urban form in supporting or impeding social economic activities. Moreover, the configurational approach can combine spatial configuration with functional mix to uncover the capacity of urban form to accommodate socio-economic activities within a certain radius. Considering both the spatial accessibility and spatial capacity could reveal the extent to which urban form could support, impede or organize socio-economic activities, that is spatial potential. Further, evaluating the spatial potential and historicity in landscape units might provide valuable information for future interventions at a finer scale (Fig. 6.28).

Spatial accessibility

This section aims to investigate whether the spatial accessibility of urban form, could reveal its locational potentials for supporting or impeding diverse socio-economic activities. The global accessibility NAIN and local integration were chosen to explore the relationship with functional diversity, as Hillier (1993) argues that more integrated locations will be more attractive as destinations for movement-reliant socio-economic activities based on the relations between street network and 'to movement'.

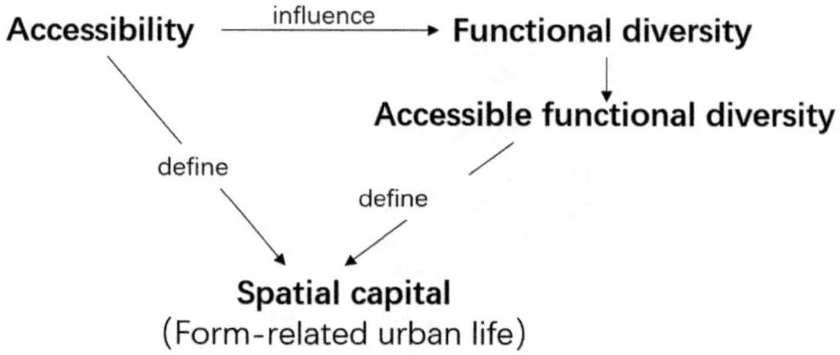

Fig. 6.28 Relationship between spatial potential, spatial accessibility and spatial capacity (*source* authors)

The socio-economic activities are measured by the functional mix of live/work/visit. The relationship between spatial accessibility and functional mix is firstly explored at the scale of physical plot, and then at the scale of 1st and 2nd tier landscape units.

At the scale of plot, its floor areas of live/work/visit functions and global/local integration were summarized and correlated in SPSS. Table 6.6 shows that the proportions of work/visit functions are positively correlated with global integration ($r = 0.225$, $r = 0.259$, $p < 0.01$) and inversely correlated with local integration ($r = -0.134$, $r = -0.257$, $p < 0.01$). This suggests that physical plots with higher proportions of work/visit functions, such as office towers, malls and museums, are likely to locate along the globally accessible streets rather than the locally accessible streets. Yet the proportion of live function correlates positively with local integration ($r = 0.309$, $p < 0.01$), and inversely correlated with global integration ($r = -0.374$, $p < 0.01$). This indicates that physical plots with higher proportion of live function are likely to aggregate within locally accessible street patchworks.

This difference shows that physical plots with different configurational properties tend to have different compositions of live/work/visit functions. The globally accessible physical plots tend to have high concentrations of work/visit functions, and higher degrees of mix, while the locally accessible physical plots tend to have higher proportions of live function, and lower degrees of mix. It could be tentatively argued that global accessibility is positively related to the degrees of functional mix and could reveal the locational potentials for diverse socio-economic activities.

For the first-order landscape units, all the floor areas of live/work/visit functions within them were aggregated (Table 6.7). Take the plan unit of the inner fringe of the Ming city wall (III), for example, the conventional MXI is 14|45|40. It is characterized by more working (45%) and visiting (40%) than living (14%). This can be largely attributed to the redevelopment of industrial sites into office, cultural and recreational facilities since 2010. For the plan unit of traditional residential settlement (II), the MXI is 86|10|4. It has bi-function of live and work. For the plan unit of market economy period settlement (V), the MXI is 60|14|27, suggesting this plan unit is mixed. It is composed of shopping malls, office towers, supermarkets and high-density residential real estate.).

Table 6.6 Correlation between global/local integration and conventional proportions of live/work/visit at the scale physical plot

		NAIN_Rn	Inte_R400	Conventional Live_per	Conventional Work_per	Conventional Visit_per
NAINl_Rn	Pearson correlation	1	−0.463**	−0.374**	0.225**	0.259**
	Sig. (2-tailed)		0.000	0.000	0.000	0.000
	N	1796	1796	1796	1796	1796
Inte_R400	Pearson correlation	−0.463**	1	0.309**	−0.134**	0.257**
	Sig. (2-tailed)	0.000		0.000	0.000	0.000
	N	1796	1796	1796	1796	1796

*Correlation is significant at the 0.05 level (2-tailed)
**Correlation is significant at the 0.01 level (2-tailed)

Figures 6.30 and 6.31 show the overlapped proportions of floor areas of live/work/visit with average global NAIN and local integration values of 1st tier plan units. At the global scale, physical plots with higher global accessibility tend to have higher proportions of work/visit functions and lower proportions of live function. At the local scale, the relationship is the opposite: physical plots with higher local accessibility tend to have lower proportions of work/visit functions and higher live function. This difference shows that urban areas with different configurational properties tend to have different compositions of live/work/visit functions. The globally accessible plan units tend to have large urban grids, high concentrations of work/visit functions, and higher degrees of mix, while the locally accessible plan units tend to have small urban grids, higher proportions of live function, and lower degrees of mix. In general, the degrees of functional mix tend to be dependent on a larger scale, and it is positively related to the global accessibility.

Nonetheless, there is one exception in that the market economy period settlement (V) has high global accessibility, yet its proportion of work/visit functions is not the highest. This is attributed to the large-scale housing estate development in the 2000s in Old Chengnan District, which overweighs the impact of spatial accessibility on functional distribution (Fig. 6.30)

Figures 6.33 and 6.34 show the overlapped proportions of floor areas of live/work/visit with average global NAIN and local integration values of 2nd tier plan units. Consistent with that in the 1st tier plan units, the degrees of functional mix of 2nd tier plan units tend to be positively related to their global accessibility and inversely related to their local accessibility. The global accessibility could reveal the locational potentials for diverse socio-economic activities.

Based on the above discussions, it could be summarized that street network structure is related to the functional mix typologies: the more accessible a landscape unit is in the city scale, the more diverse functions it has. The less accessible it is, the more mono-residential function it has. This observation is consistent with the functional implications of the dual street network structure (Hillier et al., 2007). The foreground

Table 6.7 Conventional MXI of 1st and 2nd tier landscape units

Year	Landscape unit hierarchy			Conventional MXI (live%:work%:visit%)
2015	The study area			68:13:19 Mixed
	First order	Meander of River Neiqinhuai (I)		7:44:49 Bi-functional
		Second order	a. School and institutions	12:78:9 Bi-functional
			b. Hospital	0:0:100 Mono-visit
			c. Museum	0:0:100 Mono-visit
			d. Industry	11:65:24 Mixed
		Traditional settlement (II)		86:10:4 Bi-functional
		Second order	e. Mendong-East of Zhonghua gate	92:6:3 Mono-live
			f. Menxi-West of Zhonghua gate	84:12:4 Bi-functional
		Inner fringe of Ming city wall (III)		8:52:40 Bi-functional
		Second order	g. Redeveloped industrial sites	9:50:41 Bi-functional
			h. School and institution	7:62:31 Bi-functional
			j. Vacant land	0
		Transition period settlement (IV)		84:51:11 Bi-functional
		Second order	k. Residential development since 1980	92:2:5 Mono-live
			l. Redeveloped Confucius temple area	5:1:94 Mono-visit
			m. Linear commercial development	47:27:25 Highly mixed
		Market economy period settlement (V)		60:15:27 Mixed
		Second order	n. Residential real estate	94:2:4 Mono-live
			o. Large-scale commercial development	52:6:42 Bi-functional
			p. Linear commercial development	22:50:28 Highly mixed
			q. Small-scale commercial development	41:17:42 Mixed

network of topogeometric global structure is shaped by microeconomic factors, in that socio-economic activities are likely to take advantage of the rich movement while the background network of metric local structure is largely residential, which modulates and structures movement rather than to maximize it.

Moreover, the relationship between street network structure and its functional implications could reveal the locational potentials of landscape units in supporting or impeding diverse socio-economic activities. The more globally integrated into the

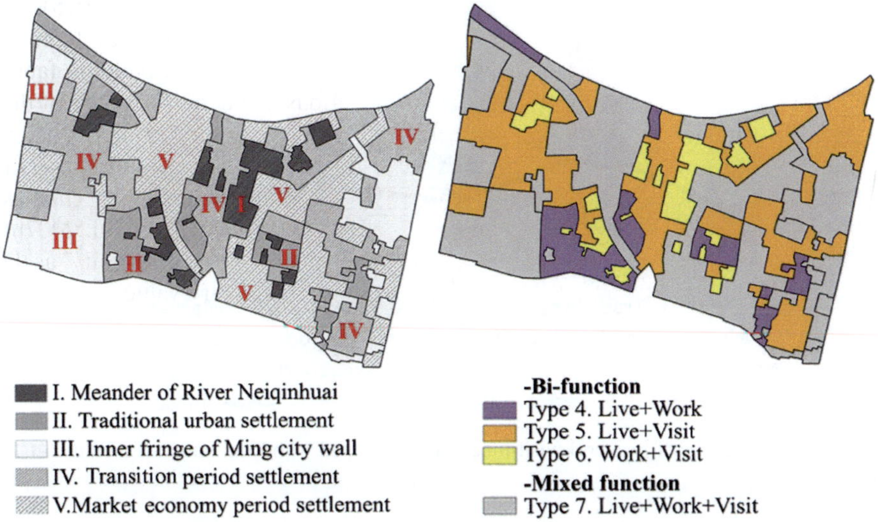

Fig. 6.29 Spatial mapping of conventional functional mix of 1st tier landscape units

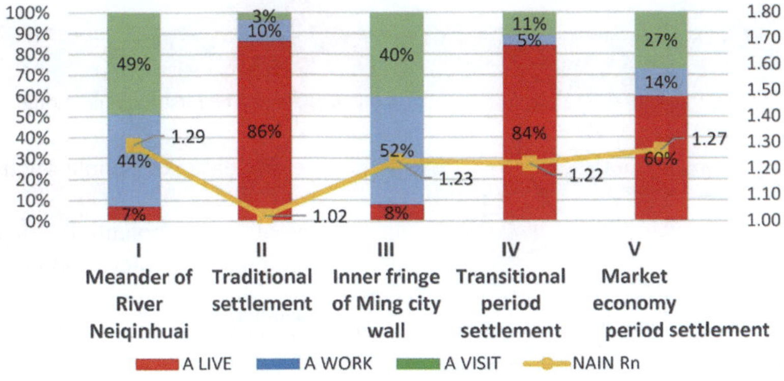

Fig. 6.30 Proportions of live/work/visit functions overlapped with global NAIN values of 1st tier plan units

city the landscape unit is, the more potentials it carries to develop socio-economic activities compared to a unit with comparatively lower integration values and hence fewer potentials.

Nonetheless, there are exceptions. Take the 1st tier landscape unit market economy period settlement (V) and the 2nd tier landscape unit Residential real estate (Vn) for example; both have high global accessibility, yet their proportion of work/visit functions is not the highest. This suggests that other factors, such as zoning, planning, or socio-economic policies may override the spatial impact on functional distribution.

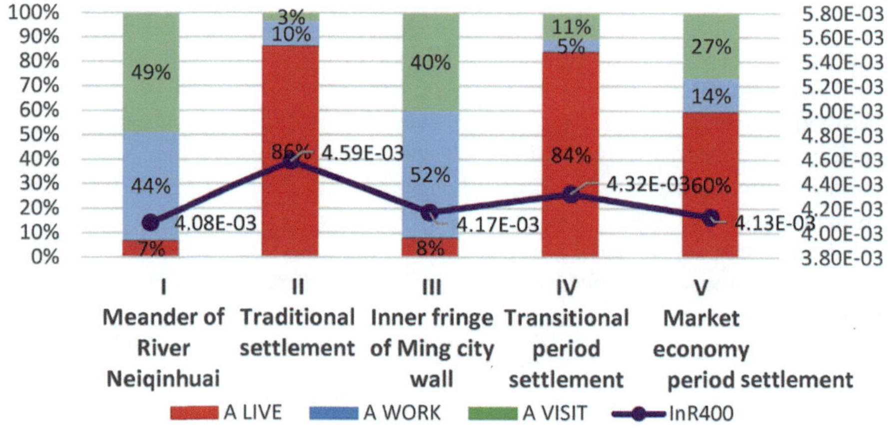

Fig. 6.31 Proportions of live/work/visit functions overlapped with local integration values of 1st tier plan units

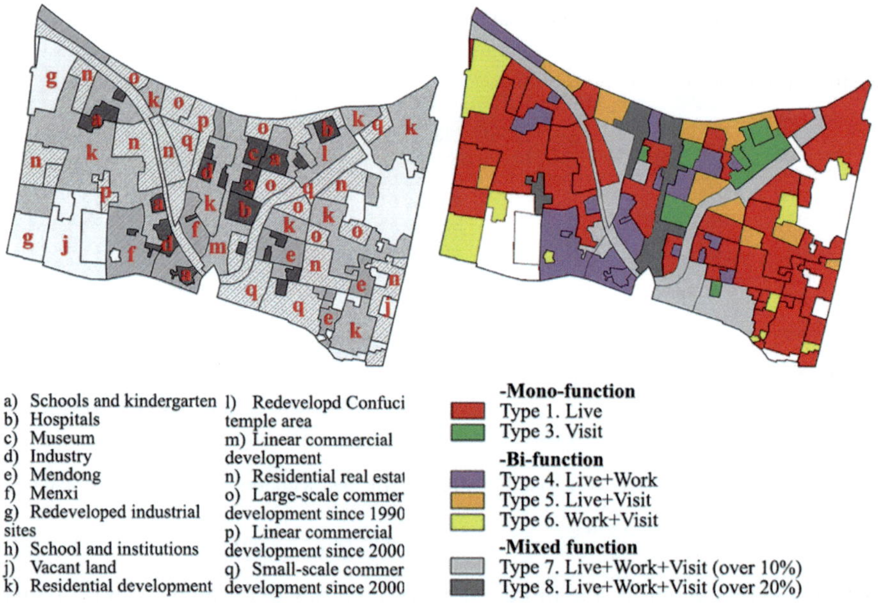

Fig. 6.32 Spatial mapping of conventional functional mix of 2nd tier landscape units

Finally, this measure of functional mix is area-based, focusing on the descriptions and analysis of a single urban element-physical plot. In reality, functions are location-based and are not limited by plot boundaries, as they could be reached within a radius from a particular location. The functional performance of a physical plot is related

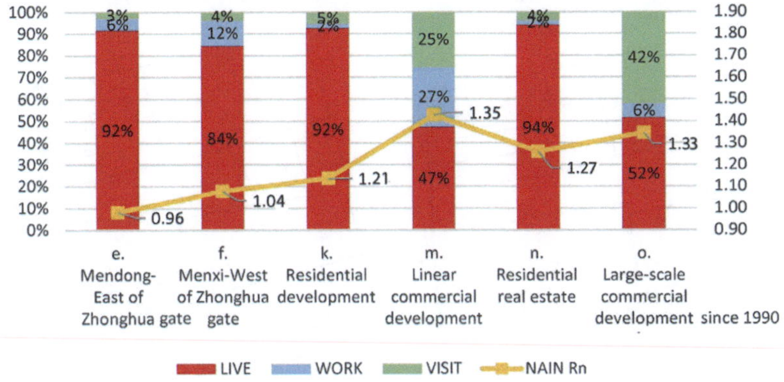

Fig. 6.33 Proportions of live/work/visit functions overlapped with global NAIN values of 2nd tier plan units

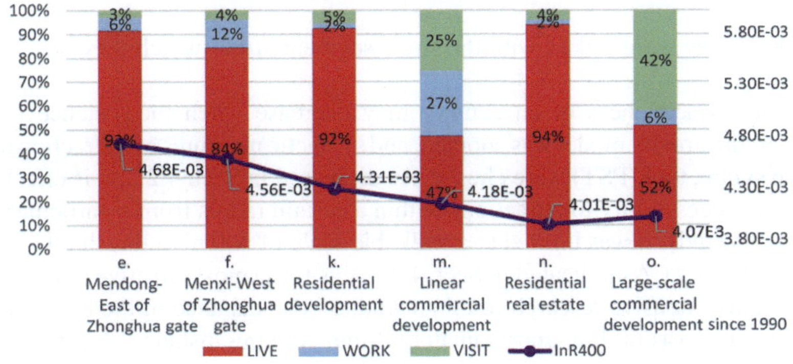

Fig. 6.34 Proportions of live/work/visit functions overlapped with local integration values of 2nd tier plan units

to its surrounding functions within a specific radius. The next section will examine this location-based functional mix using the configurational approach.

Accessible functional diversity

This section discusses the application of the configurational approach (place syntax and Urban Network Analysis toolbox) to combine the spatial analysis with functional mix to measure the accessible functional mix, which is defined in this study as the spatial capacity of urban form to accommodate socio-economic activities within a certain radius.

1. *Physical plot level*

To calculate the accessible floor areas of live/work/visit functions, in ArcGIS physical plots were input as address points with properties of floor areas of live/work/visit

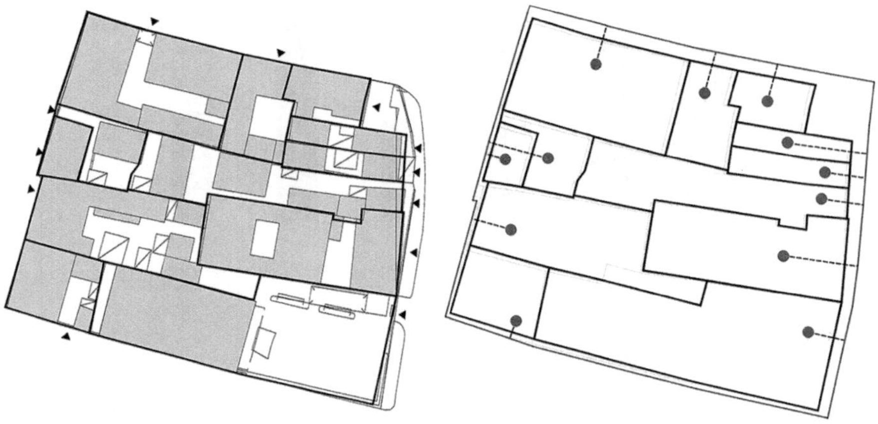

Fig. 6.35 Modelling of physical plot as address points with properties

functions. They were linked with the street segments from which they have access (Fig. 6.35).

As the attractiveness of an activity may decrease when the distance from it increases, the gravity model was adopted, and syntactic measures of 'attraction reach' in place syntax tool (PST) were chosen. As discussed in Sect. 4.2.2, it is effective in measuring accessibility to attractions within a certain radius from a particular location. In this study, it was limited to a walkable scale of 400 m (a threshold of 5-min walking distance). As indicated in Sect. 4.2.2, this is the most effective radius setting according to extensive research in urban studies. The distance of the buffer area to the study area was 600 m, approximately 8-min walking distance. A total of 1795 physical plots (study area) out of 2043 (including buffer area) were chosen, and their accessible functional mix typologies were classified in ArcGIS.

In PST, metric distance (how far to travel) and angular distance (degrees of angular shift) were used to describe relationships between line segments (dual graph). The metric distance of dual graph was visualized in Fig. 6.36, and the angular distance of dual graph is shown in Fig. 6.37. The three diagrams show a cluster pattern of accessible live/work/visit floor areas, which reveal the functional connections between adjacent plots. That is because plots take into the building floor areas of its neighbouring plots into calculation and those near the high-volume functions tend to become high. However, the pattern of angular distance is slightly different from that of metric distance. It has smaller clusters and sometimes a linear pattern of functions, because floor areas are likely to remain when the street is continuous, and decrease fast when there are directional changes.

In this study, metric distance was used to describe local residents' perceptions of distance. Since the street network in Old Chengnan District is mostly gridiron, the street direction does not change as frequently as in historic towns. People tend to choose a shorter trip length when there are fewer directional changes. Thus, accessible

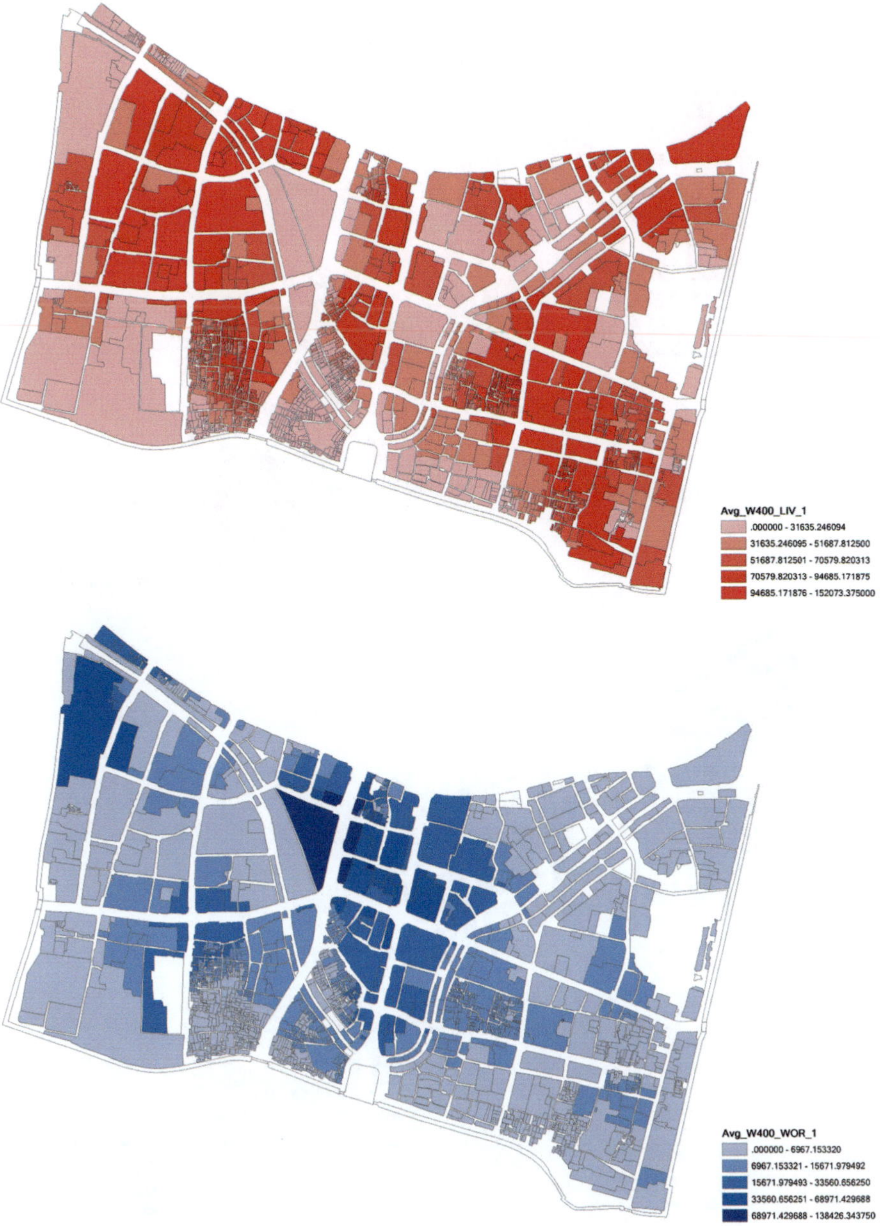

Fig. 6.36 Accessible live/work/visit floor areas in walking distance 400 m (dual graph)

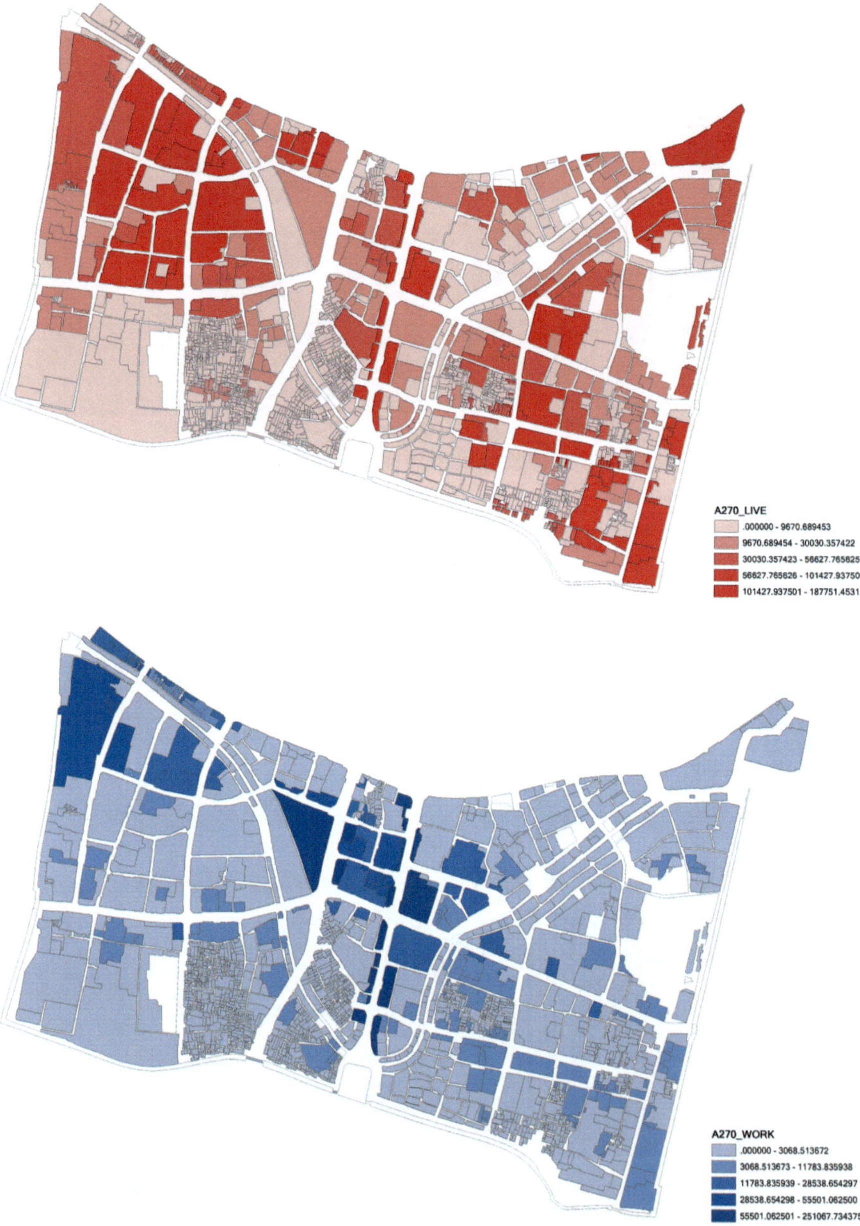

Fig. 6.37 Accessible live/work/visit floor areas in angular distance 270° (dual graph)

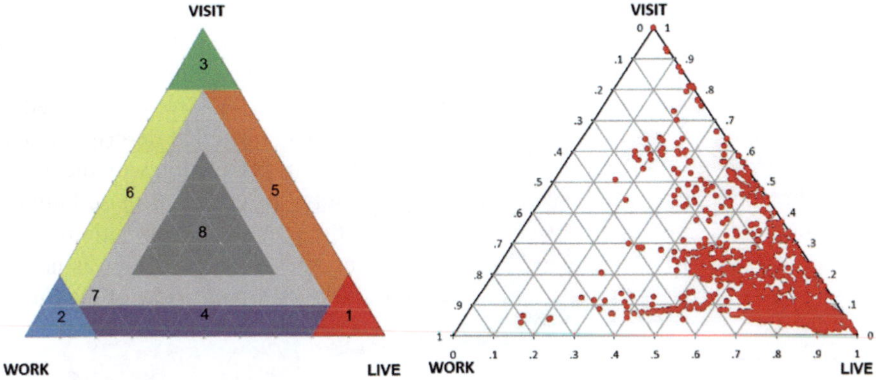

Fig. 6.38 Tertiary diagrams and distribution of physical plots based on their accessible live/work/visit percentage

function in metric distance (dual graph) was chosen as an example to further explore the functional potentials of physical plots.

Compared to the previous tertiary diagram, Fig. 6.38 shows that the number of physical plots with accessible bi-functions (live/visit and live/work) has increased greatly, and a significant cluster of physical plots with accessible mixed functions (live/work/visit) has emerged. However, the physical plots with accessible mono-visit or mono-work functions are rare, as are those with bi-functional work/visit functions.

Figure 6.39 maps the mixed typologies of each physical plot spatially. There are obvious clusters of areas with different functional mix typologies, indicating the distinctive functional potentials. Compared with the previous diagram, mono-function areas (red) are reduced greatly. Some of them have been changed into bi-functions of live/visit (orange) and live/work (purple), by taking into their neighbouring visit and work functions. The mixed-function areas have greatly increased along the Zhonghua and Zhongshan Road, where the live, work, visit functions are located in proximity and could contribute to one another.

2. *Landscape unit level*

The whole study area of Old Chengnan District has 72% accessible live, 13% accessible work and 15% accessible visit functions. The accessible MXI is 72/13/15, which suggests this district has a good balance of functions.

1st tier landscape unit

The conventional and accessible MXI of live/work/visit of 1st tier landscape units are summarized in Table 6.8 and visualized in Fig. 6.40. As can be seen in the table, traditional settlement (II) is 'bi-functional (mixed live and work)', the transitional period settlement (IV) is 'bi-functional (mixed live and visit)' and the market economy period settlement (V) is 'mixed'. The accessible mixed typologies of 1st tier plan units are all 'mixed'. This is because conventional measures capture the individual

Fig. 6.39 Mixed typology of Old Chengnan District based on their accessible floor areas of live/work/visit functions

properties of each physical plot, while accessible measures take the properties of surrounding plots into calculation.

Take the landscape unit of inner fringe of Ming city wall (III), for example, the conventional MXI is defined by more working (45%) and visiting (40%) than living (14%). This can be largely attributed to the redeveloping of the industrial sites into office, cultural and recreational facilities since 2010. The accessible MXI, however, is composed of 68% living, 15% working and 17% visiting, of which the rise in the living proportion is also attributable to the nearby residential units. This suggests the potential liveliness of these areas could be generated not only by workers and

Table 6.8 Conventional MXI and accessible MXI of 1st tier landscape units

Year	Plan unit hierarchy		Conventional MXI (live%:work%:visit%)	Accessible MXI in R 400 m (live%:work%:visit%)
2015	1st order	I. Meander of River Neiqinhuai	7:44:49 Bi-functional	58:19:23 Mixed
		II. Traditional settlement	86:10:5 Bi-functional	77:13:10 Mixed
		III. Inner fringe of Ming city wall	15:45:40 Mixed	68:15:17 Mixed
		IV. Transitional period settlement	84:5:11 Bi-functional	70:12:18 Mixed
		V. Market economy period settlement	60:15:27 Mixed	63:12:25 Mixed

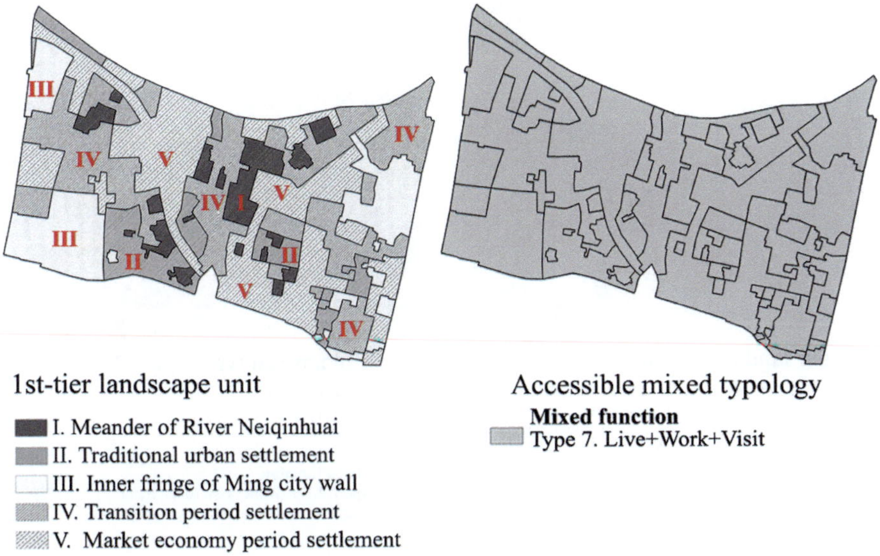

Fig. 6.40 Spatial mapping of accessible functional mix typology of 1st tier landscape units

visitors, but also by nearby inhabitants. There is great potential for socio-economic activities in this landscape unit because of the surrounding population's support.

For the landscape unit of traditional residential settlement (II), the mixed type is mono-live, in which working and visiting only take up 14% of the total floor areas. In the accessible MXI, the proportions of working and visiting have increased to 23%. This suggests that this plan unit has the potential to benefit from its surrounding working and visiting functions and become more functionally diversified. This observation also applies to the landscape unit of transitional period settlement.

For the landscape unit of market economy period settlement (V), the conventional MXI (60|14|27) and the accessible MXI (63|12|25) suggest this plan unit is mixed in both reality and potentials. It is composed of shopping malls, office towers, supermarkets and high-density residential real estate. Its surroundings play a rather limited role in this plan unit's accessible functions because the average floor areas of live/work/visit of this landscape unit are far more than that in the other landscape units.

2nd tier landscape unit

The conventional and accessible MXI of live/work/visit of 2nd tier landscape units is summarized in Table 6.9 and visualized in Fig. 6.41. This measurement reveals the functional differentiations within each 1st tier landscape unit. For example, as the conventional mixed typology of the 1st tier landscape unit of transitional period settlement (IV) is bi-functional (mixed live and visit). Yet its 2nd tier landscape units have three different mixed typologies: plan unit of residential development (k) is

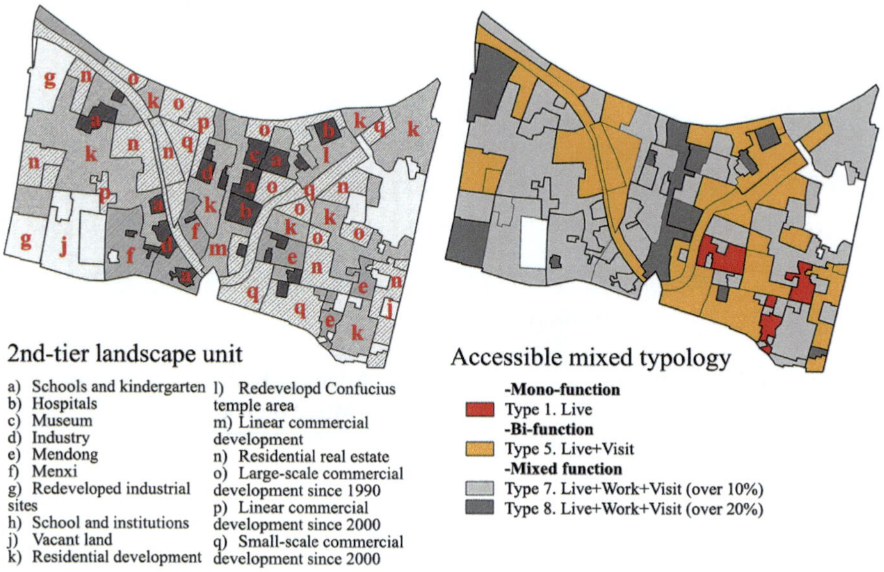

Fig. 6.41 Spatial mapping of accessible functional mix typology of 2nd tier landscape units

mono-live, the linear commercial development (*m*) is mixed and the redeveloped Confucius temple area (*l*) is mono-visit.

To sum up, the configurational approach (PST) has been combined with live/work/visit functions in measuring the accessible functional mix (spatial capacity). This moves one step further from conventional geographic descriptions, by emphasizing the functional connections between adjacent plots. It has provided an enhanced understanding of urban areas, by treating them as interrelated entities instead of isolated patches. Meanwhile, this spatial capacity reveals the ability of urban space to accommodate diverse activities accessible to a location at different hierarchies of landscape units. Nonetheless, some areas with high spatial capacity are less integrated with the rest of the city. This disadvantage of location may impede further development of socio-economic activities in these areas. Thus, spatial capacity alone is not enough to capture the potentials of urban space to engender diverse activities. In the next section, the combination of spatial accessibility with spatial capacity together to investigate how urban form could support, impede and organize urban life, will be examined.

Spatial potential for generating socio-economic vibrancy

As discussed in Sect. 4.2.2, if areas have high accessible functional mix, this indicates that they have higher spatial capacities to engender diverse activities. If areas have high global accessibility, this suggests they have higher locational potentials to support socio-economic activities. Combining these two spatial metrics creates the opportunity to propose a new classification system of various types of spatial potential, and to divide urban areas into clusters with varied degrees of spatial potentials. This helps to identify which spatial metric has low values, what kinds of improvements are needed to the spatial or functional features at finer scales of urban areas.

In this classification system, physical plots are firstly classified into three types based on their global integration values. As discussed in Sect. 4.2.2, global integration is positively related to the degrees of functional diversity. They are divided in ArcGIS by using the natural break method into high, medium and low values, underlying the different degrees of locational potentials to support socio-economic activities. Type 1 represents the high global NAIN value, which indicates high locational potentials to locate work/visit functions in the city. Type 2 represents medium global NAIN, which suggests in-between locational advantage. Type 3 represents the low global NAIN, which suggests poor locational advantage (Fig. 6.42). Secondly, functional mix typology is classified into high, medium and low values. Mono-function is defined as low value, bi-function is medium value, and mixed function is high value (Fig. 6.43).

The combination of spatial accessibility and accessible functional mix generates three categories: 'matched', 'in-between' and 'mismatched' (Table 6.10). 'Matched' typology reflects similar values in accessible functional mix and global accessibility,

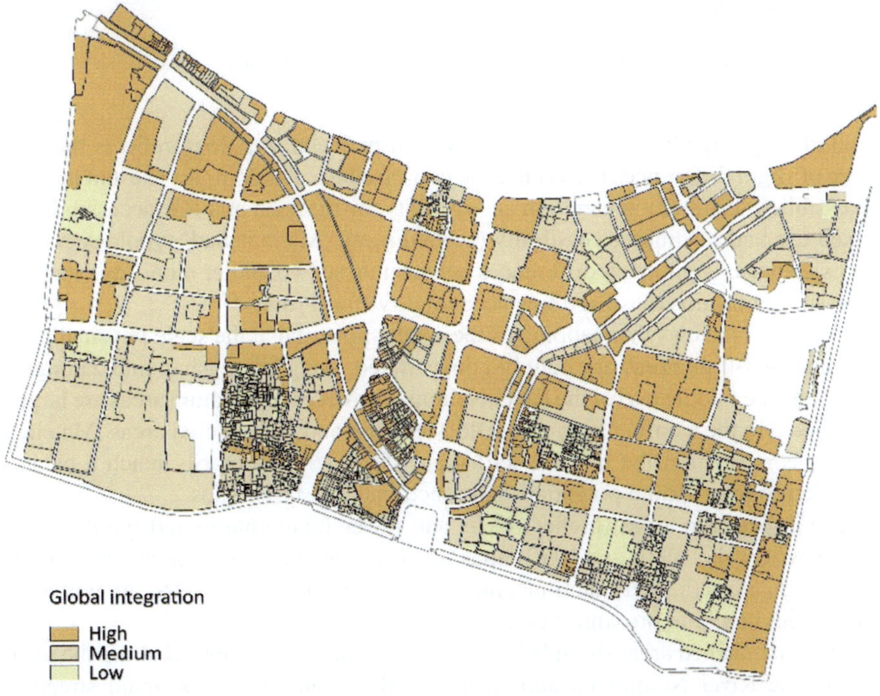

Fig. 6.42 Classification of physical plots based on their spatial accessibility

Fig. 6.43 Classification of physical plots based on their spatial capacity

similarly high, medium or low (first row). 'In-between' typology reflects the existence of slight differences between these two variables, either mono-use and medium accessibility, bi-function and high accessibility, bi-function and low accessibility or mixed use and medium accessibility (second row). 'Mismatched' typology reflects the existence of significant differences between these two, either mono-use and high accessibility or mixed use and low accessibility (third row).

Figure 6.44 shows the mapping of the spatial potential at the scale physical plots. The matched-high areas are mainly concentrated along Zhongshan and Zhonghua Street and the middle part of the River Neiqinhuai. Matched-medium areas are largely distributed in the Confucius temple, Old Mendong and New Menxi areas. Matched-low areas are within the historic settlement. They all temporarily reach a balance considering their spatial potentials and functional mix typologies.

The in-between areas (higher NAIN and lower functional mixed typology) are largely within the 1980s and 1990s housing units, together with Gongyuan area. The in-between areas (higher functional mixed typology and lower NAIN) are within the large street block, in proximity to major streets.

The mismatched areas (high NAIN and mono-use) are distributed along the lower part of the River Neiqinhuai and Jiangning Road, and along the main streets of the Hehuatang historic area. The mismatched areas (low NAIN and mixed use) are located within the historic area Diaoyu Tai.

Table 6.10 Classification of spatial potentials into matched, in-between and mismatched based on the accessible functional mix typology and global NAIN

A	B	A + B
Spatial capacity	Spatial accessibility	Spatial potential
High (mixed)	High	Matched high
Medium (bi-function)	Medium	Matched medium
Low (mono)	Low	Matched low
Low (mono)	Medium	In-between (HNAIN, LMIX)
Medium (bi-function)	High	
Medium (bi-function)	Low	In-between (HMIX, LNAIN)
High (mixed)	Medium	
Low (mono)	High	Mismatched (HNAIN, LMIX)
High (mixed)	Low	Mismatched (HMIX, LNAIN)

Fig. 6.44 Mapping of spatial potential at the scale physical plot

This analytical framework provides a clear description of the variations of spatial potential at the scale of physical plot. It presents the opportunities to discern clusters of urban areas with same spatial potential. Based on this classification, this framework suggests ways of improving the functional performance of urban areas at finer scales. For example, area with 'mismatched' (low accessibility and high functional diversity) spatial potential is suggested to improve its global accessibility, increasing its locational potentials for maintaining and developing its functional diversity. Nonetheless, such a measurement does imply that the higher accessibility and spatial capacity, the higher spatial potential, but it does not imply a higher spatial potential is always better. Indeed, higher spatial potential may indicate higher possibilities of changes which may threaten the continuity of historicity.

6.3.3 Combined Analysis

This section aims to evaluate the strength and weakness of landscape units by combining historicity and spatial potential. Figure 6.45 shows the landscape unit division in the market economy period. The historical expressiveness of urban form is uncovered by the hierarchical mapping of landscape units. Together with a detailed written description and photographs (Table 6.11), it provides evidence to identify what is most important to the historic character and to evaluate whether the building block, plot and street pattern are compatible with its surroundings in the same unit. Meanwhile, the spatial potentials of the 1st, 2nd and 3rd tier landscape units are summarized, uncovering how urban form could support or impede diverse socio-economic activities. Combining both historicity and spatial potential evaluates paves way for an overall understanding of the strength and weakness of landscape units before zooming into the 4th tier landscape units.

Figure 6.46 shows the overlapped mapping of landscape unit and spatial potential. Within the 4th tier landscape units, there are clusters of physical plots (5th tier) characterized with varied degrees of spatial potentials, such as matched high, in-between or mismatched. This subdivision of 4th tier landscape units provides a finer-scale analysis of the landscape units in terms of historicity and spatial potential, which might help build up management priorities of urban landscape with regard to what to maintain and what to change, and provide reference for future small-scale and incremental interventions.

The landscape unit of the inner fringe of the Ming city wall (III) is taken as an example to elaborate how this analytical framework could uncover the risks and challenges of landscape units. This plan unit dates back to the Ming Dynasty and has undergone a lot of changes driven by dynamic forces in the economic, social and cultural spheres. At first, it mainly consisted of farmland, open space and vacant land. After 1949, the state-led industrialization located large- and medium-sized factories there. Since the 1990s, some of the factories were converted into housing, and some into creative industrial parks under the policy of industrial restructuring and real estate development. The increasing global syntactic values suggest that this

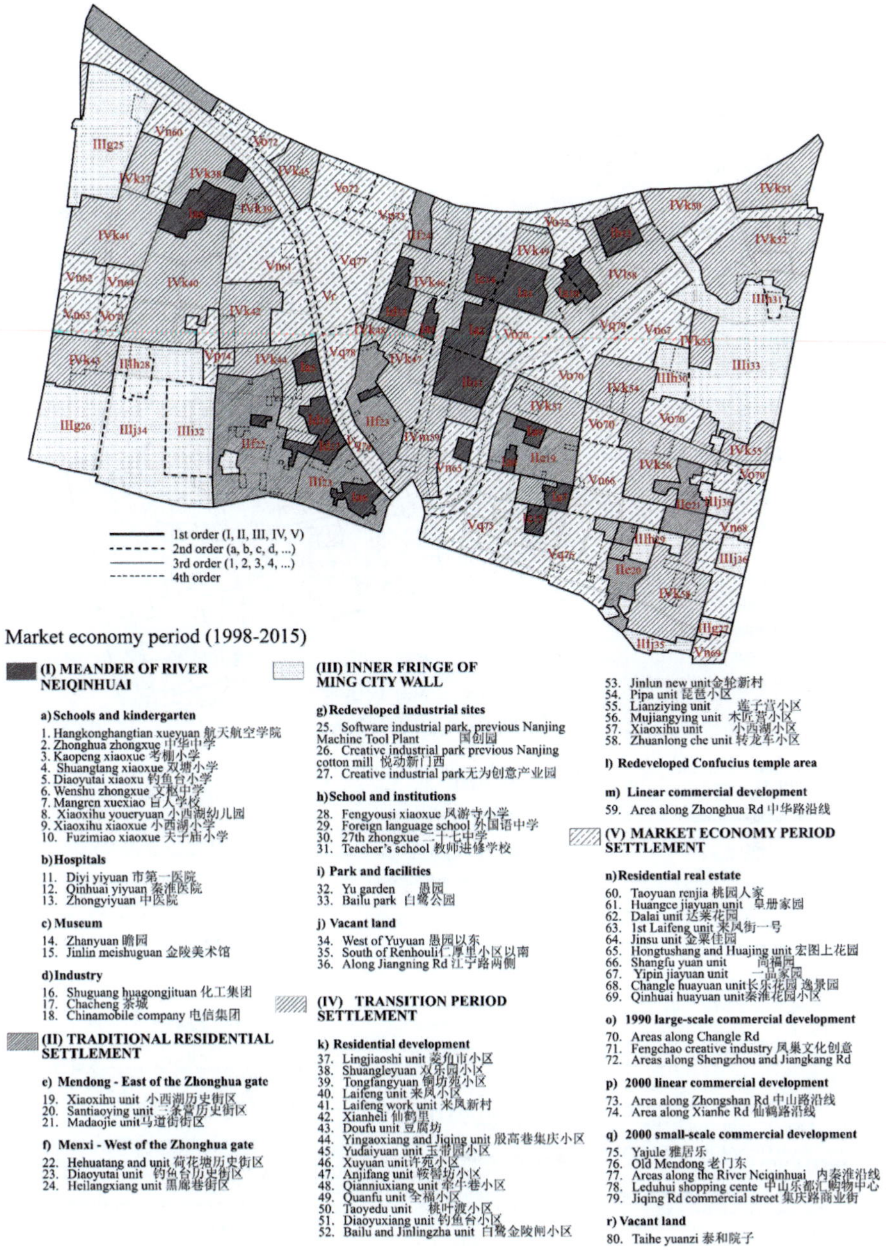

Fig. 6.45 Landscape unit of Old Chengnan District

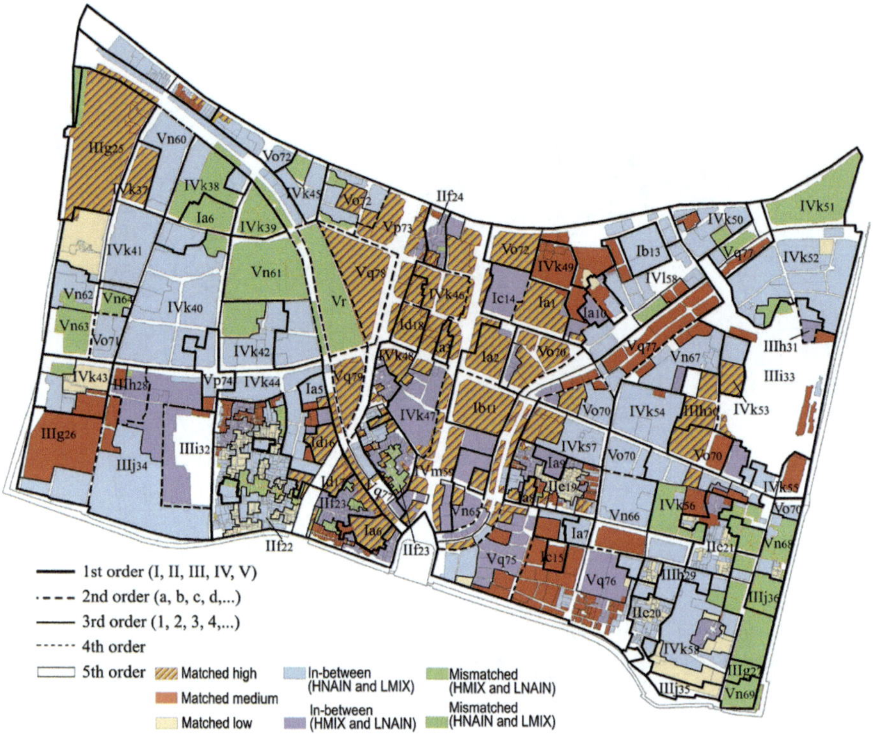

(I) MEANDER OF RIVER NEIQINHUAI

a) Schools and kindergarten

1. Hangkonghangtian xueyuan 航天航空学院
2. Zhonghua zhongxue 中华中学
3. Kaopeng xiaoxue 考棚小学
4. Shuangtang xiaoxue 双塘小学
5. Diaoyutai xiaoxu 钓鱼台小学
6. Wenshu zhongxue 文枢中学
7. Mangren xuexiao 盲人学校
8. Xiaoxihu youeryuan 小西湖幼儿园
9. Xiaoxihu xiaoxue 小西湖小学
10. Fuzimiao xiaoxue 夫子庙小学

b) Hospitals

11. Diyi yiyuan 市第一医院
12. Qinhuai yiyuan 秦淮医院
13. Zhongyiyuan 中医院

c) Museum

14. Zhanyuan 瞻园
15. Jinlin meishuguan 金陵美术馆

d) Industry

16. Shuguang huagongjituan 曙光化工集团
17. Chacheng 茶城
18. Chinamobile company 电信集团

(II) TRADITIONAL RESIDENTIAL SETTLEMENT

e) Mendong - East of the Zhonghua gate

19. Xiaoxihu unit 小西湖历史街区
20. Santiaoying unit 三条营历史街区
21. Madaojie unit 马道街街区

f) Menxi - West of the Zhonghua gate

22. Hehuatang and unit 荷花塘历史街区
23. Diaoyutai unit 钓鱼台历史街区
24. Heilangxiang unit 黑廊巷街区

(III) INNER FRINGE OF MING CITY WALL

g) Redeveloped industrial sites

25. Software industrial park, previous Nanjing Machine Tool Plant 国创园
26. Creative industrial park previous Nanjing cotton mill 悦动新门西
27. Creative industrial park 无为创意产业园

h) School and institutions

28. Fengyousi xiaoxue 风游寺小学
29. Foreign language school 外国语中学
30. 27th zhongxue 二十七中学
31. Teacher's school 教师进修学校

i) Park and facilities

32. Yu garden 愚园
33. Bailu park 白鹭公园

j) Vacant land

34. West of Yuyuan 愚园以东
35. South of Renhouli 仁厚里小区以南
36. Along Jiangning Rd 江宁路两侧

(IV) TRANSITION PERIOD SETTLEMENT

k) Residential development

37. Lingjiaoshi unit 菱角市小区
38. Shuangleyuan 双乐园小区
39. Tongfangyuan 铜坊苑小区
40. Laifeng unit 来凤小区
41. Laifeng work unit 来凤新村
42. Xianheli 仙鹤里
43. Doufu unit 豆腐坊
44. Yingaoxiang and Jiqing unit 殷高巷集庆小区
45. Yudaiyuan unit 玉带园小区
46. Xuyuan unit 许苑小区
47. Anjifang unit 鞍辔坊小区
48. Qianniuxiang unit 牵牛巷小区
49. Quanfu unit 全福小区
50. Taoyedu unit 桃叶渡小区
51. Diaoyuxiang unit 钓鱼台小区
52. Bailu and Jinlingzha unit 白鹭金陵闸小区
53. Jinlun new unit 金轮新村
54. Pipa unit 琵琶小区
55. Lianziying unit 莲子营小区
56. Mujiangying unit 木匠营小区
57. Xiaoxihu unit 小西湖小区
58. Zhuanlong che unit 转龙车小区

l) Redeveloped Confucius temple area

m) Linear commercial development

59. Area along Zhonghua Rd 中华路沿线

(V) MARKET ECONOMY PERIOD SETTLEMENT

n) Residential real estate

60. Taoyuan renjia 桃园人家
61. Huangce jiayuan unit 皇册家园
62. Dalai unit 达莱花园
63. 1st Laifeng unit 来凤街一号
64. Jinsu unit 金粟佳园
65. Hongtushang and Huajing unit 宏图上花园
66. Shangfu yuan unit 尚福园
67. Yipin jiayuan unit 一品家园
68. Changle huayuan unit 长乐花园 逸景园
69. Qinhuai huayuan unit 秦淮花园小区

o) 1990 large-scale commercial development

70. Areas along Changle Rd
71. Fengchao creative industry 风巢文化创意
72. Areas along Shengzhou and Jiangkang Rd

p) 2000 linear commercial development

73. Area along Zhongshan Rd 中山路沿线
74. Area along Xianhe Rd 仙鹤路沿线

q) 2000 small-scale commercial development

75. Yajule 雅居乐
76. Old Mendong 老门东
77. Areas along the River Neiqinhuai 内秦淮沿线
78. Leduhui shopping cente 中山乐都汇购物中心
79. Jiqing Rd commercial street 集庆路商业街

r) Vacant land

80. Taihe yuanzi 泰和院子

Fig. 6.46 Overlapped mappings of spatial potential with landscape unit of Old Chengnan District plan unit has become more spatially embedded in the urban structure as the city has grown, which provides opportunities for locating socio-economic activities in this plan unit. The relatively low local syntactic values are attributed to the sparse street network and large, irregular urban plots. This landscape unit's conventional MXI is 15|45|40, in which the proportion of work function is the highest among all the 1st tier plan units, due to the schools and upgraded industrial parks. Its accessible MXI is 68|15|17, indicating its potentials to engender balanced socio-economic activities by taking the surrounding residential population into consideration. Its high global accessibility also supports this functional potential.

Within the inner fringe of the Ming city wall (III), one of the 2nd tier plan units is the redeveloped industrial sites (IIIg). They are the products of modern industrialization and have high heritage values. Since the 2000s, a lot of industrial sites have been revamped for creative companies and start-ups, such as the Software Industrial Park and the Culture and Creative Industrial Park. Their global integration value is high, providing opportunities for functional upgrade in these areas in this area. Its conventional MXI is 9|50|41, and accessible MXI is 53|26|20, indicating that these industrial sites have good potentials to become functional diverse with the support of nearby residential population.

In the redeveloped industrial sites (IIIg), there are three 3rd tier landscape units: IIIg25, IIIg26 and IIIg 27. IIIg25 is the Nanjing Software Industrial Park, recently updated from previous Nanjing Machine Tool Plant built in the 1950s, which occupies 8 ha of land. It was designed with large-scale warehouses in row layout (Fig. 6.47). It is spatially well connected to the rest of the city. Its conventional MXI is 0|87|13 and accessible MXI is 29|59|12, indicating that this site has potential balanced functions. The spatial potential is 'matched high' because the high spatial accessibility could support and maintain its potential mixed functional diversity. This suggests no interventions are currently needed for its balanced relationship between locational and functional potentials.

The landscape unit of the transitional period settlement (IV) is taken as another example. It has emerged since the economic reform initiated in 1979, when modern residential units consisting of slab apartments and grid streets began to be built, as well as the large-scale commercial and business redevelopment along the major

Fig. 6.47 Bird's eye view and entrance view of redeveloped Nanjing software park (IIIg25) (*source* http://www.kanshangjie.com/article/60365-1.html)

axial street. Its global accessibility is medium. Its conventional MXI is 84|5|11, in which the residential function is the second highest among all the 1st tier landscape units. Its accessible MXI is 70|12|18, indicating the potential of this landscape unit to condition balanced socio-economic activities by taking surrounding work and visit functions into consideration. The medium global accessibility on one hand controls the movement flows and on the other hand supports this functional potential.

Residential development (IVk) is one of the 2nd tier landscape units. It is characterized with multi-storey slab building blocks dating back to the 1980s. It has a variety of retail businesses and services on the ground floor, creating a vibrant walking experience for local residents (Fig. 6.48). The global syntactic value is medium. Its conventional MXI is 92|2|6 and accessible MXI is 76|10|14, suggesting these residential units have good potentials to become functionally diverse with the support of nearby work and visit functions.

Residential development (IVk) is further divided into five typologies with varied spatial vitality. The first has 'in-between' spatial potential, composed of mono-live and medium accessibility. It is the major type of spatial potential in IVk. It is suggested that more work or visit functions could be added to take advantage of its medium spatial accessibility.

The second (IVk47) has 'in-between' spatial potential, composed of mixed typology and medium accessibility. It indicates that the spatial accessibility may not fully support its mixed typology. Its spatial accessibility could be improved in the future.

The third (IVk54) has 'in-between' spatial potential, composed of bi-functional typology (live + visit) and high accessibility. It is suggested more work function could be developed to reach a mixed functional balance, taking advantage of its high spatial accessibility.

The fourth (IVk46 and IVk53) has 'matched' high spatial potential, which indicates that it is well connected to the rest of the city. Its mixed function ensures the area's vitality 24/7.

The fifth (IVk38, IVk39, IVk51 and IVk56) has 'mismatched' spatial potential, composed of mono-live and high accessibility. It is suggested that more visit and work functions could be added to fully take advantage of its high global accessibility.

Fig. 6.48 Street view of multi-storey residential buildings (IVk) (*source:* authors)

These two examples have exemplified how the strength and weakness of landscape units are evaluated by the integrated approach in terms of historicity and spatial potential. It shall be noted that the high spatial potential is not always better, as it may threaten the historic character of urban areas. A balanced view is required based on the different planning and conservation objective and scenarios. For example, the 1st and 2nd tier landscape units of the inner fringe of Ming city wall (III) have matched spatial potential, indicating its spatial advantage to support diverse socio-economic activities, while at the same time it should be noted that the government has a role to oversee to avoid over commercialization which might undermine its industrial heritage value.

6.4 Discussion

This chapter has applied the analytical framework that combines the historico-geographical approach and the configurational approach in investigating the historical development and spatial potential of urban form by using Old Chengnan District in Nanjing as an example. The first part explored how the configurational approach could uncover the generative rules of urban growth which are related to plan unit formation and transformation. The syntactic patterns and values could substantiate plan unit recognition to some extent. The historico-geographical approach provided a place-specific context to interpret the underlying social logic of spatial configuration. The key findings are summarized at the end of Sect. 6.2.3.

The second part described the analytical framework that could evaluate the strength and weakness of landscape units by overlapped mappings and descriptions of their historicity and spatial potential. On one hand, it examines what is important to the historic character and whether existing urban forms accord with the character of urban landscape units to which they belong; on the other hand, it identifies spatial metrics with low values, for example, poor spatial locations which may not support the diverse content or areas with single use which may not fully take advantage of its locational advantage. This evaluation has uncovered the strengths and weaknesses of urban landscape units, which could provide reference for urban planners and designers to manage change, particularly what new forms to incorporate into the old urban landscape, and what the locational and functional choices of new development to make.

Spatial potential in this study refers to how urban form could support, impede or organize vibrant socio-economic activities. The configurational approach could contribute to the understanding of the spatial potential in landscape units at finer scales in terms of spatial accessibility and spatial capacity. The spatial accessibility was measured by global integration, which reveals the locational potentials of urban form in supporting or impeding diverse socio-economic activities. The spatial capacity was measured by configurational analysis with land use data and emerging open data Point of Interest (POI) data. It uncovers the capacity of urban form to accommodate diverse socio-economic activities within a certain walking distance.

How this evaluation could inform future landscape management, and add efforts to reconciling the conflicts between conservation and development? This will be explored further by two historic urban areas in Old Chengnan District of Nanjing in Chap. 7.

Table 6.11 Evaluation of different hierarchies of landscape units

Landscape unit (1st, 2nd, 3rd, 4th-tier)				Socio-economic vibrancy		Historic character			Evaluations of strength and weaknesses of landscape units
				Spatial potential		Ground plan	Building block	Land use	
				Accessible functional mix	Spatial accessibility				
I. Meander of River Neiqinhuai				Mixed 58:19:23	High 1.291	This unit has the large and medium-sized plots (average size 5942 m^2) along the River Neiqinhuai, which date back to the Feudal period. The large institutional buildings are the major type			This landscape unit has a long history. Some of the buildings are listed historic buildings. Both its functional mix potential and spatial accessibility are high
2nd order	a. School and institutions			Mixed 64:17:19	Medium 1.233	Large urban grid with regular plot shape	Mid-rise building in strip block	Institutional function has remained since the transition period	The functional consistency is attributed to Nanjing's long established administrative and cultural identity
	3rd-order	1, 2, 3, 4, 6, 8		Matched high					These areas are well connected to the rest of the city. The mixed function ensures the vitality of the area 24/7
				Mixed	High				
		5		In-between					These areas have good functional mix potentials (either bi or mixed). Their good spatial locations in the city could support future functional redevelopment • 5 is near the business cluster along Xianhe Rd and has rich work function potentials • 7 is in the proximity of Old Mendong and Confucius temple area, and it has rich visit function potentials • 9 and 10 the medium and low spatial accessibility may not support their functional diversity
				Live + work	High				
		7		In-between					
				Live + work	High				
		9		In-between					
				Mixed	Medium				
		10		In-between					
				Live+work	Low				

(continued)

Table 6.11 (continued)

	b. Hospital		Highly mixed 31:25:43	High 1.700	Large urban grid with regular plot shape	Large building block built in the transition period	Hospital function has remained since the transition period	These areas are well connected to the rest of the city. The mixed function ensures the vitality 24/7
	3rd-order	11, 12, 13	Matched high					
			Mixed	High				
	c. Museum		Highly mixed 29:29:42	Medium 1.167	Large urban grid with regular plot shape	Traditional courtyard building	They are redeveloped from traditional housing in the transition period and market economy period separately	Ic14 is national heritage, consisting of a group of historic buildings
	3rd-order	14	In-between					The medium global accessibility may not support its current functional mix potential
			Highly mixed	Medium				
		15	Matched medium			Large building block		The medium global accessibility could support its bi-functional (live/visit) potentials
			Live + visit	Medium				
	d. Industry		Mixed 59:28:13	High 1.478	Large urban grid with irregular plot shape	Large building block built in the market economy period	Industry function emerged in the planned economy period	Some of the large and medium-sized buildings are representative of Nanjing's industrialization
	3rd-order	16, 18	Matched high					These areas are well connected to the rest of the city. The mixed function ensures the vitality 24/7. The spatial accessibility and spatial capacity support each other
			Mixed	High				
		17	Matched high					
			Mixed	High				

(continued)

Table 6.11 (continued)

II. Traditional settlement			Mixed 77:13:10		Low 1.022	This unit is composed of small-sized plots (average size 227 m^2) with narrow frontage and deep depth. It is divided by Zhonghua Rd into Mendong and Menxi. The street pattern is cul-de-sacs. Courtyard housing is the prevalent building type			This unit has high historical and cultural values. There is a number of listed historic buildings. Its low global accessibility may not support its high functional potential
2nd order	e. Mendong		Mono live 82:9:9		Low 0.9623	Cul-de-sac street alleyways with small and rectangular plot shape	Courtyard housing dating back to the Feudal period and planned economy period	Residential function could be traced back to the Feudal period	They are the dilapidated residential settlements dating back to the Ming and Qing dynasties. The street alleyways and building fabric mark Nanjing's Old Chengnan historic character
	f. Menxi		Mixed 74:15:11		Low 1.039				
	3rd-order	19, 20, 21, 22, 23, 24	4th-order	Matched low					• The low global accessibility could restrict the movement flows in the residential areas • The medium global accessibility could support its live/work functional potentials • These areas are well connected to the rest of the city. The mixed function ensures the vitality 24/7
				Mono live	Low				
				Matched medium					
				Live + work	Medium				
				Matched high					
				Mixed	High				
				In-between					• The low street network structure may not support its current functional mix potential • The global accessibility has not been fully used. These areas could be updated for more other use
				Live + work	Low				
				In-between					
				Mono live	Medium				
				Mismatched					The global accessibility has not been fully used. These areas could be updated for mixed use
				Mono live	High				

(continued)

Table 6.11 (continued)

III. Inner fringe of Ming city wall			Mixed 68:15:17	Medium 1.232	This unit has large-sized plots (average size 9153 m^2), which are formed since the Feudal period. Industrial buildings are the major type			This plan unit has a long history with rich industrial heritage. Its globally accessibility provides a good spatial advantage for functional mix potentials. Its functional mix potential is also high
2nd order	g. Redeveloped industrial sites		Highly mixed 53:26:20	Medium 1.254	Large urban grid with large-sized plots	Large building block dating back to the planned economy and early modern period	IIIg25 was redeveloped from previous Nanjing Machine Plant, and IIIg26 from Nanjing Textile factory, in the market economy period	They are the cultural icons of Nanjing's modern industrialization
	3rd order	25	Matched high					• These areas are well connected to the rest of the city. The mixed function ensures the vitality 24/7 • The medium global accessibility could support its live/visit functional potentials
			Mixed	High				
		26	Matched medium					
			Live + visit	Medium				
		27	Mismatched					The global accessibility has not been fully used. These areas could be updated for more mixed use
			Mono-live	High				
	h. School and institution		Mixed 75:12:13	Low 1.079	Large urban grid with medium-sized plots	Row building dating back to the planned economy period	They are redeveloped from previous public buildings or vacant land in the early modern and modern period	They are located in the adjacency of residential settlement, contributing to the daytime vitality within these areas
	3rd-order	28	In-between					• The medium global accessibility may not support its current functional mix • The global accessibility has not been fully used. These areas could be updated for more other use
			Highly mixed	Medium				
		29	In-between					
			Mono live	Medium				
		30	Matched high					These areas are well connected to the rest of the city. The mixed function ensures the vitality 24/7
			Mixed	High				
		31	In-between					The global accessibility has not been fully used. These areas could be updated for more mixed use
			Live + work	High				

(continued)

Table 6.11 (continued)

IV. Transition period settlement			Mixed 70:12:18		Medium 1.217	This unit has medium-sized plots (average size 4266 m^2), which mainly resulted from plot amalgamation dating back to the 1980s. There is a mixture of modernist slab apartments and medium-sized commercial buildings			This plan unit is characterized with pseudo-traditional and modernist building styles. Its globally accessibility provides a good spatial advantage for functional mix potentials. Its functional mix potential is also high
2nd order	k. Residential development		Mixed 76:10:14		Medium 1.214	Medium-grid street network with medium-sized plots	Slab apartments emerged during transition period	They are redeveloped from traditional housing	Landscape unit 'k' has a variety of retails and services on the ground floor, creating a vibrant walking experience
	3rd-order	37, 40, 41, 42,43,44,45, 50, 52, 57, 58	4th-order	In-between					The global accessibility has not been fully used. These areas could be updated for more other use
				Mono live	Medium				
		46, 53		Matched high					These areas are well connected to the rest of the city. The mixed function ensures the vitality 24/7
				Mixed	High				
		47		In-between					The medium global accessibility may not support its current functional mix potential
				Mixed	Medium				
		38, 39, 51, 56		Mismatched					The global accessibility has not been fully used. These areas could be updated for mixed use
				Mono live	High				
		54		In-between					The global accessibility has not been fully used. These areas could be updated for mixed use
				Live + visit	High				

(continued)

Table 6.11 (continued)

	l. Redeveloped Confucius temple		Bi-functional 38:3:60		Low 1.037	Small-grid street network with small-sized plots	Pseudo-traditional building	It was previously a worship and imperial examination place in the Feudal period. The temple was rebuilt in the 1980s	It has been the cultural and tourist icon of Nanjing
	3rd-order	58	4th-order	In-between					The global accessibility has not been fully used. These areas could be updated for mixed use
				Live + visit	High				
				In-between					
				Mono-visit	Medium				
				Matched medium					The medium global accessibility could support its live/visit functional potentials
				Live + visit	Medium				
				Mismatched					The global accessibility has not been fully used. These areas could be updated for mixed use
				Mono-visit	High				
	m. Linear commercial development		Highly mixed 49:25:26		Medium 1.349	Along the north-south axis with medium-sized plots	Medium building block	They are redeveloped from traditional housing	It is characterized by medium commercial building blocks in 'qilou' style
	3rd-order	59	Matched high						These areas are well connected to the rest of the city. The mixed function ensures the vitality 24/7
			Mixed		High				

(continued)

Table 6.11 (continued)

V. Market economy period settlement			Mixed 63:12:26		High 1.271	This unit has medium-sized plots (average size 2850 m^2). There is a mixture of modernist slab buildings, towers, and large-scale commercial buildings dating back to the late 1990s			This plan unit is characterized with real estate housing and commercial development. Its high global accessibility provides good spatial advantage for its functional mix. Its functional mix potential is also high
2nd order	n. Residential real estate		Bi-functional 81:7:12		Medium 1.274	Large-grid street network with medium-sized plots	Slab buildings and tower building	They are redeveloped from traditional housing	Landscape unit 'n' is enclosed with controlled access. It is low in work and visit functions
	3rd-order	60, 67	4th-order	In-between					The global accessibility has not been fully used. These areas could be updated for more other use
				Mono live	Medium				
		61, 62, 63, 64, 68, 69		Mismatched					The global accessibility has not been fully used. These areas could be updated for mixed use
				Mono live	High				
		65		In-between					The medium global accessibility may not support its current functional mix potential
				Mixed	Medium				
		66		In-between					The global accessibility has not been fully used. These areas could be updated for mixed use
				Live + visit	High				
				Matched medium					The medium global accessibility could support its live/visit functional potentials
				Live + visit	Medium				

(continued)

Table 6.11 (continued)

	o. Large-scale commercial development		Bi-functional 59:9:32		Medium 1.331	Large-grid street network with large-sized plots	Large building block	They are redeveloped from traditional housing and transition period housing	Plan unit 'o' has large urban grids, making it less friendly for pedestrians in the neighbourhood scale. It is also low in work function
	3rd-order	70, 72	4th-order	Matched high					These areas are well connected to the rest of the city. The mixed function ensures the vitality 24/7
				Mixed	High				
				In-between					The global accessibility has not been fully used. These areas could be updated for mixed use
				Live + visit	High				
		71		In-between					
				Live + work	High				
	p. Linear commercial development		Mixed 41:39:19		Medium 1.462	Large-grid street network with medium-sized plots	Tower building	They are redeveloped from traditional housing	Plan unit 'p' is well balanced, both globally and locally accessible
	3rd-order	73	4th-order	Matched high					These areas are well connected to the rest of the city. The mixed function ensures the vitality 24/7
				Mixed	High				
		74		In-between					The global accessibility has not been fully used. These areas could be updated for mixed use
				Live + work	High				

(continued)

Table 6.11 (continued)

	q. Small-scale commercial development		Bi-functional 63:9:28		Medium 1.223	Small street alley-ways with medium-sized plot	Pseudo-traditional building	They are redeveloped from traditional housing	Plan unit 'q' is both globally and locally accessible. It is low in work function
	3rd-order	75, 76	4th-order	In-between					• The global accessibility has not been fully used. These areas could be updated for more other use • The low global accessibility may not support its current functional mix potential
				Mono live	Medium				
				In-between					
				Live + visit	Low				
				Matched medium					• The medium global accessibility could support its live/visit functional potentials • These areas are well connected to the rest of the city. The mixed function ensures the vitality 24/7
				Live + visit	Medium				
		77		Matched high					
				Mixed	High				The global accessibility has not been fully used. These areas could be updated for mixed use
				In-between					
				Live + visit	High				
				Matched medium					• The medium global accessibility could support its live/visit functional potentials • These areas are well connected to the rest of the city. The mixed function ensures the vitality 24/7
				Live + visit	Medium				
		78		Matched high					
		79		Mixed	High				
		80		Mismatched					The global accessibility has not been fully used. These areas could be updated for mixed use
				Mono live	High				

References

Bandarin, F., & Van Oers, R. (2012). *The historic urban landscape: Managing heritage in an urban century*. Wiley.

Chen, F., & Thwaites, K. (2013). Chinese urban design: the typomorphological approach. Farnham: Ashgate Pub. Ltd.

Chen, Y. (2017). *Jinling gujin tukao* [*The historical maps of Nanjing*]. Nanjing Chubanshe (in Chinese).

Conzen, M. R. G. (1960). Alnwick, Northumberland: a study in town-plan analysis. Transactions and Papers (Institute of British Geographers), iii-122.

Conzen, M. R. G. (1973). Geography and townscape conservation. Paper presented at the Anglo-German Symposium in Applied Geography, Giessen-Würzburg-München.

Fang, Z., Hua, X., Liu, Q., Xu, Y., Du, L., & Chen, W. (2012). Jiucheng baohu de zaisikao [Reflections on the conservation of the historic city of Nanjing]. *Shijie Jianzhu Daobao, 2*, 10–33. (in Chinese).

Han, P. (2004). *Qinhuaihe shihua* [*History of Qinhuai river*]. Nanjing chubanshe.

He, L., & Cui, G. (2000). Nanjing Chengshi Kongjian Kuozhande tezheng yu jizhi [Characteristic and mechanism of Nanjing's spatial growth]. *Chengshi Guihua 6*, 56–60 (in Chinese).

Hillier, B., Penn, A., Hanson, J., Grajewski, T., & Xu, J. (1993). Natural movement: or, configuration and attraction in urban pedestrian movement. Environment and Planning B: Planning and Design, 20(1), 29–66.

Hillier, B. (1996). *Space is the machine: A configurational theory of architecture*. Cambridge University Press.

Hillier, B. (1996). Cities as Movement Economies. Urban Design 1(1), 41–41.

Hoek, J. v. d. (2010). The mixed use index (MXI) as planning tool for (new) towns in the 21st century. In M. Provoost (Ed.), *New towns for the 21st century: The planned versus the unplanned city*. Martien de Vletter, SUN.

Jacobs, J. (1961). *The death and life of great American cities*. Wiley.

Koh, M., & Melic, K. (2017). *A historic heart how heritage districts can make cities more liveable*. Retrieved from Singapore: https://www.clc.gov.sg/docs/default-source/reports/a-historic-heart.pdf

Kropf, K. (1997). Typological zoning. Typological process and design theory, 127–140.

Liu, X., & Long, Y. (2015). Automated identification and characterization of parcels with OpenStreetMap and points of interest. *Environment and Planning B: Planning and Design, 43*(2), 341–360.

Liu, Y., & Wu, F. (2006). Urban poverty neighbourhoods: Typology and spatial concentration under China's market transition, a case study of Nanjing. *Geoforum, 37*(4), 610–626.

Nanjing Local Chronicle Committee. (2008). *Nanjing Chengshiguihuazhi* [*The urban planning history of Nanjing*]. Jiangsu renmin chubanshe (in Chinese).

Scheer, B. (2010). *The evolution of urban form: Typology for planners and architects*. Routledge.

Shan, J. (2006). Cong daguimo jiufang gaizao dao xunxi jianjin youji gengxin [From mass demolish to incremental redevelopment]. *Wenwu 7*, 26–40.

Wang, L., & Mo, C. (2017). Chengshi gengxin he fengmao baohu de chengshi sheji yu chengshi zhili shijian [Urban design and governance for urban renewal and landscape preservation]. *Guihuashi, 33*(10), 135–141. (in Chinese).

Whitehand, J., & Gu, K. (2017). Urban fringe belts: Evidence from China. *Environment and Planning B: Urban Analytics and City Science, 44*(1), 80–99.

Wu, L. (1994). *Beijing Jiucheng yu jier hutong* [*The old city of Beijing and Juer Hutong*]. Zhongguo jianzhu gongye chubanshe (in Chinese).

Xie, Y., & Costa, F. J. (1991). Urban design practice in socialist China. *Third World Planning Review, 13*(3), 277.

Xue, B. (2015). *Nanjing Chengshi shi* [*The history of Nanjing city*]. Dongnan daxue chubanshe (in Chinese).

Yang, Z., & Chen, J. (2014). Housing reform and the housing market in urban China. In *Housing affordability and housing policy in urban China* (pp. 15–43). Springer Berlin Heidelberg.
Ye, N., Tan, Z., & Li, H. (2014). *Nanjing Bainian Chengshishi 1912–2012* [*A hundred-year historicy of Nanjing 1912–2012*]. Nanjing chubanshe (in Chinese).
Zhang, Y., Liu, Q., & Zhang, J. (2011). Lishi chengqu chiduxia de lishididuan xitong baohu zhengzhi gongzuofangfa tantao [The study of the conservation strategies of historic urban areas-based on the conservation planning of Old Chengnan]. *Chengshi guihua lunwenji*, 8295–8305 (in Chinese).

Part III
Implementation and Reflections

Chapter 7
Urban Conservation and Management Practice in China

Abstract Since the economic reform, the rapid urbanization of Chinese cities has greatly speeded up the demolition of historic urban fabric in the old cities. The relationship between conservation and development has become contradictory and competing. In recent years, urban planning in China aims to optimize the existing built environment with features of 'small-scale and incremental' urban renewal and heritage conservation. Nanjing as a typical historic city is chosen to elaborate the evaluation and management framework proposed by the local planning authority. Two historic urban areas, Hehuatang and Diaoyutai, are taken as an example to introduce how this framework could help evaluate building values, build up their hierarchical order and generate different conservation and redevelopment strategies accordingly.

Keywords Urban conservation and redevelopment · Historic urban area · Nanjing · Conservation policy and practice

7.1 Background

This chapter firstly describes the phenomenon and evolution of area-based conservation, which is a common trend for the conservation policy and practices in many Chinese cities. Then it zooms in on Nanjing, elaborating in detail its recent area-based conservation guidelines, with a focus on two historic urban areas: Hehuatang and Diaoyutai. The chapter ends with several conclusions on the strengths and weaknesses of the local planning authority's assessment framework and management proposal of these two urban areas.

Since the economic reform in 1978, China has experienced rapid commodification of land during the urbanization process. A series of land reforms have been introduced since the 1980s to allow purchase and transfer of the land utility rights (Wu, 2002). This sparked an explosion of real estate developments in Chinese cities. New economic pursuit, design of new urban landscapes and real estate housing development all took place at the same time. This push for change has greatly threatened urban heritage. To preserve the historic identity of cities, central government has launched conservation laws and regulations. A list of Chinese Precious Historic and

X. Li and Y. Zhang, *Conserving and Managing Historical Urban Landscape*, The Urban Book Series, https://doi.org/10.1007/978-981-19-4222-8_7

Cultural Cities (lishi wenhua mingcheng) was issued in 1982, and the concept of Historic and Cultural Conservation Areas (HCCAs) (*Lishi Wenhua Baohuqu*) was introduced by the State Council in 1986 (Ruan & Sun, 2001). The Historic and Cultural Conservation area consists of street blocks, groups of buildings, and even entire towns and villages with traditional characteristics representative of certain historic periods. Conservation planning is to be based on a thorough investigation of the area's historical character. It aims to maintain and enhance its local historical character and coordinate conservation and economic growth. Nonetheless, the principles are too loose to regulate and guide conservation projects, especially on what to keep and what to change in implementation (Chen, 2011).

Since the 1990s, to maximize the economic potential of inner-city land, the historic inner cities have become the focus of urban development and regeneration processes, which led to the demolition of old buildings, the relocation of populations and the erection of new buildings. 'Massive demolition and massive construction' was the common practice (Ruan, 2004). Take one historic area Yanliaofang in Nanjing for example; it is located near the River Neiqinhuai and occupies 7.1 ha of land (Fig. 7.1). It was previously characterized by traditional courtyard housing dating back to the Ming and Qing dynasties. In 2006, the local government launched a redevelopment project in Old Chengnan District to improve the city's image, raise revenues and pursue political achievement. Buildings were demolished regardless of Historic and Cultural values. Urban plots were amalgamated into two sites with different uses: site A was taken over by Huarun Enterprise for high-end commercial redevelopment; site B was taken by the Taihe Enterprise for luxury courtyard housing development. The new construction has no relations to the historical tissue, neither to what has been demolished nor to the remaining surroundings.

Another example is Xintiandi of Shanghai, which marks a turning point in the government's policy from massive demolition to conservation (Chen, 2011). It covers an area of approximately 3 ha (Fig. 7.2). It used to be a traditional residential area

Fig. 7.1 Left: previous urban fabric of Yanliaofang historic area. Right: Redevelopment plan and birds' eye view of Yanliaofang historic area. *Source* sina.cn

Fig. 7.2 Left: aerial view of original two blocks of redevelopment at Xintiandi. *Source* Shui on Land Ltd; Right: Street view of buildings after renovation at Xintiandi. *Source* google streetmap

characterized with buildings in 'Shikumen' architectural style which is a combination of Western and Chinese traditional features. Xintiandi project was implemented by a property developer (Shui on Land Ltd) from Hong Kong in 1996. The architectural elements and facades of the 'Shikumen' buildings were preserved, while the functions were updated entirely. This could be regarded as an effort to conserve the buildings' historical and cultural values and turn them into economic benefits. The project was opened for business in 2002 and Xintiandi has become a high-end commercial and tourist venue, consisting of high-end coffees, clubs, restaurants and boutiques (Zhu & Goethert, 2010). It achieved great commercial success by raising the property values in this area, and this model has been copied in other Chinese cities.

Compared to the first case, in the second one conservation planning in redevelopment has made an effort to restore the historical character and surroundings and to respond to the transformation of government's policies from the 1980s. Nonetheless both examples are typical property-led projects carried out in historic cities under the gradual intensification of the market-oriented economy. There are a growing number of criticism and reflections on this practice, not only because the physical fabric as the container of historic identity and memory no longer exists as well as the irreversibly undermined historical values and architectural values of old cities (Ruan, 2001; Zhong & Chen, 2017), but also because the continuity of urban life in historic urban areas is hardly concerned and respected (Chen, 2011). A careful and sensitive planning and management of the economic resource is emphasized so as not be in conflict with the concern to protect and conserve the Historic and Cultural values (Fang et al., 2012).

Researchers have long argued that historic districts could serve as powerful catalysts for economic and social regeneration (Bandarin & Van Oers, 2012; Jacobs, 1961; Koh & Melic, 2017). Through careful small-scale, incremental modifications of form and function as a source for changes, the development of diverse economic activities and an improvement in the quality of life could be achieved (Fang et al., 2012; Shan, 2006; Wang & Mo, 2017; Wu, 1994; Zhang et al., 2011). Since the 2010s, there has been a transformation of Chinese government's planning and design policy that intends to bring down the coarse level of city development to smaller scale of city regeneration, to ensure the long-term sustainability of conservation planning, to maintain historicity and improve socio-economic viability (MOHURD, 2018). Nonetheless, this shift is more reflected in the planning policy than in the realized plans currently. As in practice, local governments, planners and designers and different stakeholders are all responsible for such socio-economic sustainability in the conservation and regeneration of historical urban areas. In this chapter, our focus is on the urban planner and designer's expertise in preparing the conservation plan, as well as its strengths and limitations.

In the conservation plan of Historic and Cultural Conservation Areas, methods and strategies are largely derived from conventional city planning, like designated conservation area boundaries, preserved physical forms and proposed land use changes. Nonetheless, many fundamental problems remain unsolved, such as the boundary delimitation and criteria establishment of conservation priorities (Whitehand, 2007). One of the key consequences is that in the management proposals ill-informed geographical boundaries of historic area usually cut across landscape units, which are products of a city's historical development. The other key consequence is that management priority is largely focused on individual buildings, ignoring the relationship between historic building and the urban area they are part of, which often results in a fragmented urban landscape. This has largely limited the role of conservation plan in regulating and guiding the urban conservation and redevelopment practice.

7.2 Area-Based Conservation in Nanjing, China

7.2.1 Evaluation and Management Framework

Nanjing developed its first *Conservation Planning of Precious Historic and Cultural Cities* in 1984. The loose conservation guidelines and rapid economic development have limited effective conservation of the historic heritage in the rapid urbanization process. Large-scale demolition of historic urban fabric was still prevalent in Old Chengnan District in the 2000s. To balance the conflicts between conservation and development, the *Guidelines on Conservation Planning of Precious Historic and Cultural City of Nanjing* was published in 2010. In the Guidelines, conservation principles for 'overall conservation, public sector-led and private sector-participated, and small-scale redevelopment' were put forward for the first time. The major concern was twofold: physical form and utilization. Historic urban fabric should be conserved as much as possible and the major living function of historic areas should not be changed. These principles became the guidelines for preparing the latest version of *Conservation Planning of Precious Historic and Cultural Cities (2010–2030)* published in 2012. In this latest version, a systematic framework to evaluate the rich and diverse historic resources before developing the management proposals was built up by the planning authority. The evaluation is at three scales: city, area and building (Liu & Shen, 2010; Zhou, 2010).

In the city scale, three historic districts are identified: Old Chengnan District (6.9 km^2), Ming Palace District (6.5 km^2) and Drum Tower-Qingliang Mountain District (6.2 km^2) (Fig. 7.3). Among them, Old Chengnan District is characterized as a traditional settlement dating back to the Ming and Qing dynasties.

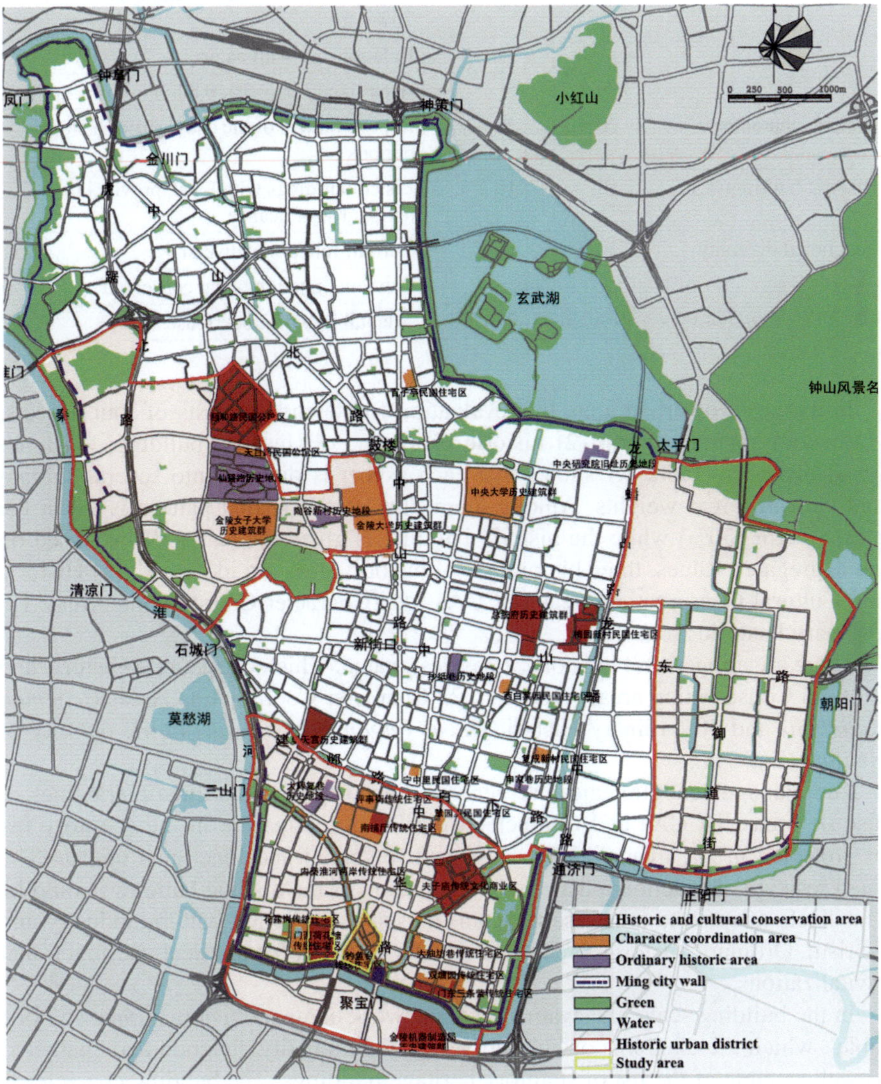

Fig. 7.3 Locations of three types of historic areas in the 2012 Conservation Planning proposal

Table 7.1 Evaluation framework of historic area (Zhou, 2010)

Aspects	Weights (%)		Details
Overall spatial pattern	50	15	Relationship with the city
		10	Continuous urban fabric
		15	Proportions of historic buildings (60%, 40%, 20%)
		5	Relationship with nature
		5	Area size (3 ha, 1 ha, 0.8 ha)
Historic character	20	20	Representative of the character of historic periods
Historic building	10	10	Listed buildings with local, provincial or national significance
Functional diversity	20	10	Current functional diversity
		3	Previous function being conserved
		7	Adaptability of modern use

In the scale of urban area, this evaluation framework consists of four aspects: (1) overall spatial pattern; (2) historic character; (3) historic building; (4) functional diversity. Table 7.1 shows that each aspect is divided into several subfactors with different weights. Among them, the overall spatial pattern is given the most weight (50%) while the historic building is given the least (10%). Based on the aggregated values, three hierarchies of historic areas are identified: (1) Historic and Cultural Conservation areas (HCCAs); (2) Character Coordination Areas; (3) Ordinary Historic Area.

There are altogether 41 historic areas designated, which consist of 9 Historic and Cultural Conservation areas (HCCAs), 22 Character Coordination Areas (*fengmao xietiaoqu*) and 10 Ordinary Historic Areas (*yiban lishi diduan*) in the old city of Nanjing (Fig. 7.3).

The management strategies are generated according to the three hierarchies of historic areas: HCCAs and Character Coordination Area are in the statutory list of protected urban areas which need sensitive spatial and functional update. The main residential function must remain, and some of the historic buildings could be sensitively adapted to other uses, such as museums, galleries, cultural industry and tourism. Ordinary Historic Areas have greater freedom for physical and functional revitalizations.

In the building scale, the evaluation framework contains 'five diagrams and one table', which are diagrams of building age, height, quality, function and character, as well as table of the list of statutory heritage buildings. The building character is evaluated by its significance in defining the unique character of different historic periods. These diagrams are at a finer scale of 1:500–1:1000.

Table 7.2 shows that the evaluation covers three aspects: building age, character and quality. Four hierarchies of buildings are identified: heritage building, historic building, character building and ordinary building. Heritage buildings are defined as listed historic buildings with local, provincial or national significance. Historic buildings are those with a long history, which could well represent the character of a particular historic period. Character buildings are historic buildings with less representativeness of a particular period or buildings built after 1949 with distinctive features. Ordinary buildings are either historic buildings with the least representativeness or buildings built after the 1949 but in poor condition. This evaluation framework contributes to the delimitation of area boundaries and establishment of conservation hierarchy within historic urban areas (Zhou, 2010).

The management strategies are generated based on the hierarchy of building values: (1) maintenance, (2) repair, (3) renovation, (4) alteration and demolition (Table 7.3). Maintenance is to tackle the deterioration of heritage and historic buildings. Repair is to improve the building structure and remedy damage in historic and character buildings. Renovation is to coordinate the building facade with its surroundings and adaptively reuse the character and ordinary buildings. Alteration and demolition apply to those buildings whose character is incompatible with their surroundings and which are in poor condition (Zhang, 2012).

Table 7.2 Evaluation framework of building values (Zhang, 2012)

Aspect	Building age		Before 1949	After 1949		All periods	
	Building character		1st, 2nd rank	3rd rank	1st, 2nd rank	4th rank	All ranks
	Building quality		1st, 2nd, 3rd rank				4th rank
Evaluation		Heritage building	Historic building	Character building		Ordinary building	

Table 7.3 Conservation and redevelopment strategy of buildings (Zhang, 2012)

Building age		Ming, Qing, and Republican period			After 1949	All periods	
Building character		1st, 2nd rank		3rd rank	1st, 2nd rank	4th rank	All ranks
Building quality		1st, 2nd	3rd	3rd	1st, 2nd	1st, 2nd	4th rank
Evaluation	Heritage building	Historic building		Character building		Ordinary building	
Strategies	Maintenance		Repair		Renovation		Alteration; demolition
	Keep the building in good order and limit its further deterioration		Remedy defects by damage or use. Minor adaptations but not involving alteration or renovation		Structural changes to interior space Accord building character with its surroundings		Control the building height. Coordinate the outlooks with its surroundings

This latest version of conservation planning has built up a multi-hierarchy framework, which ranges from the city scale to the building scale. Evaluations of historic values are developed at each scale and serve as a basis for developing corresponding management strategies in a more objective and evidence-based way compared to previous versions of conservation planning.

7.2.2 Evaluation and Management of Two Historic Urban Areas

In this section, two historic urban areas—Hehuatang and Diaoyutai—are taken as examples to elaborate on the planning authority's evaluation framework and management proposal. They are located to the west of Zhonghua gate in Old Chengnan District. This is the densely populated historic district in Nanjing; the average population density of Old Chengnan District had reached 41,000 people per km^2 in 2010, according to the Sixth National Population Census then (Shen et al., 2013).

As mentioned previously, a large number of historic areas in Old Chengnan District have been demolished in the rapid urban development of Nanjing. There are only a few isolated patches left, including Hehuatang and Diaoyutai areas. The total size of the two areas is around 25 ha. They are characterized by courtyard housing dating back to the Ming and Qing dynasties and used to be the most prosperous mixed residential and commercial settlement of Nanjing. These areas gradually decayed with the decline of traditional handicraft industry at the beginning of the twentieth century (Han, 2004). Buildings are gradually becoming dilapidated and are in a poor condition. The living facilities which are needed for modern urban life have not been appropriately developed. Historic resources are surrounded by deteriorated and impoverished urban fabric which no longer functions in the traditional way.

According to the evaluation framework at the area scale, Hehuatang is defined as Historic and Cultural Conservation area (HCCA), and Diaoyutai as a Character Coordination Area in the *Conservation Planning of Precious Historic and Cultural City of Nanjing (2010–2030)*. The Conservation proposals for these two areas were prepared by Nanjing's planning authority in 2012. The main objective was to strike a balance between conservation and development: firstly, to maintain the integrity of urban landscape and preserve dilapidated historic buildings; secondly, to provide new economic activity to replace or compliment indigenous uses which are in decline or have disappeared from the area; thirdly, to control the overcrowded population by reducing the number of households to half and to improve the living standards of local residents. The first part of the planning document includes a detailed evaluation of buildings (at the scale of 1:500–1:1000) within the historic urban areas. The second part includes proposed physical and functional interventions for these two areas based on the evaluations.

The evaluation framework examined five building attributes: building character, age, quality, height and function. The building character (Fig. 7.4) was mainly in

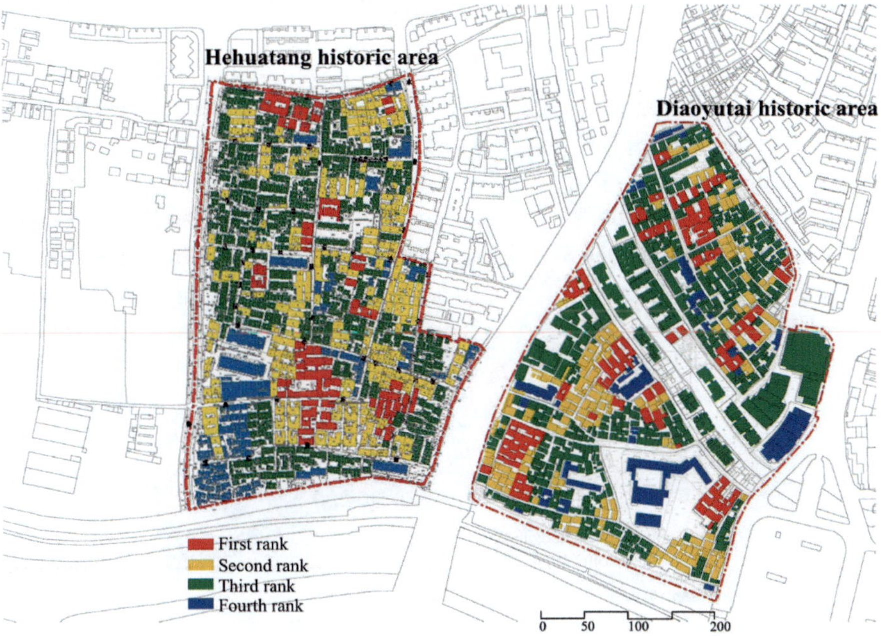

Fig. 7.4 First attribute: building character hierarchy

the third rank, that is the historic buildings had aesthetic and symbolic values which could represent the distinctive character of the Ming and Qing dynasties in terms of building scale, structure, materials.

With regard to age, around 50% of buildings dated back to the Feudal period and the Republican period, while the rest were built after 1959 (Fig. 7.5).

The building quality (Fig. 7.6) was mainly in the third rank, indicating the building structure was partly damaged yet could be restored.

The majority of buildings were one or two-storeys high, except for some three- or four-storey buildings (Fig. 7.7).

The major building function was residential (Fig. 7.8). There were some mixed uses scattered along Yinma alleyway, and some large institutional plots and commercial plots along Zhonghua and Zhongshan Road.

By evaluating the above five attributes, the planning authority divided buildings into four hierarchies: heritage buildings, historic buildings, character buildings and ordinary buildings (Table 7.2). As shown in Fig. 7.9, the heritage and historic buildings constituted 29% of the total buildings, character buildings 42% and ordinary buildings 29%. This hierarchical order of buildings was to help generate different conservation strategies and set up a basis for physical and functional revitalization at the building scale.

Based on the hierarchical order of buildings, different interventions were generated in the conservation proposal. Figure 7.10 shows the five hierarchies: buildings to

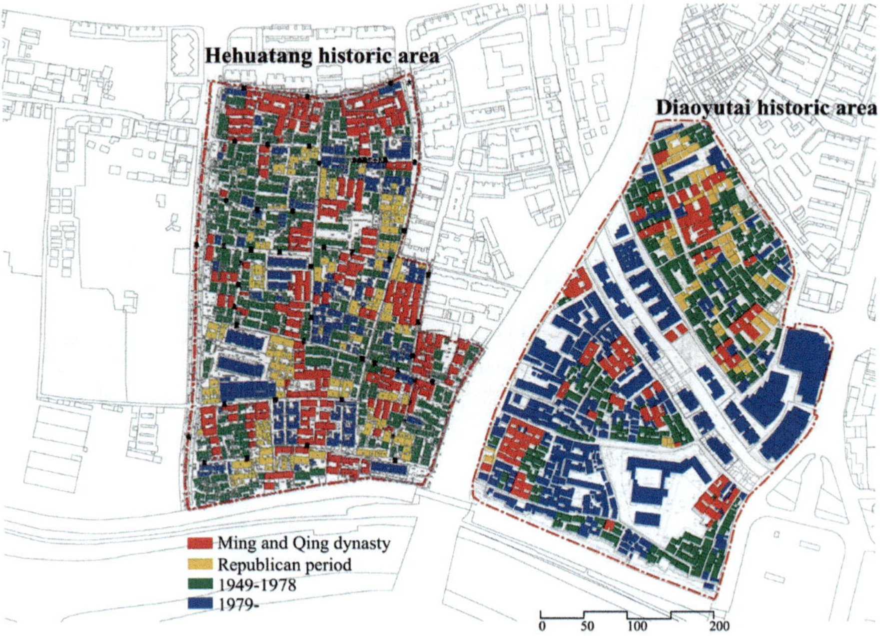

Fig. 7.5 Second attribute: building age

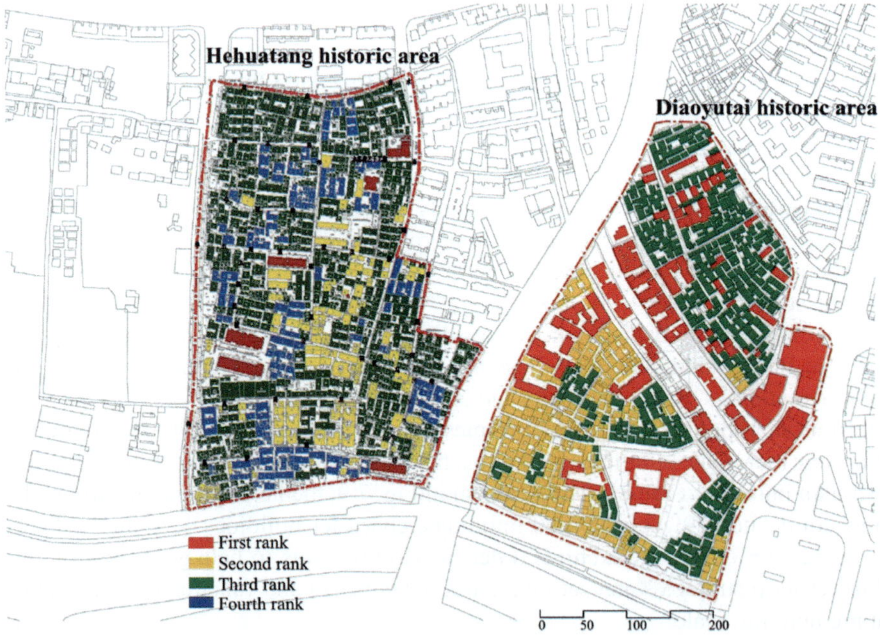

Fig. 7.6 Third attribute: building quality

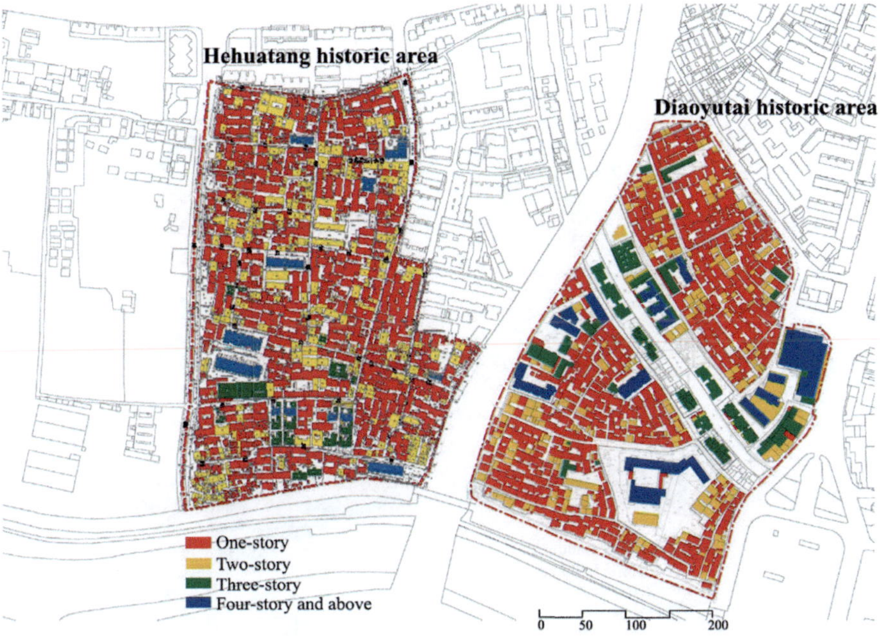

Fig. 7.7 Fourth attribute: building height

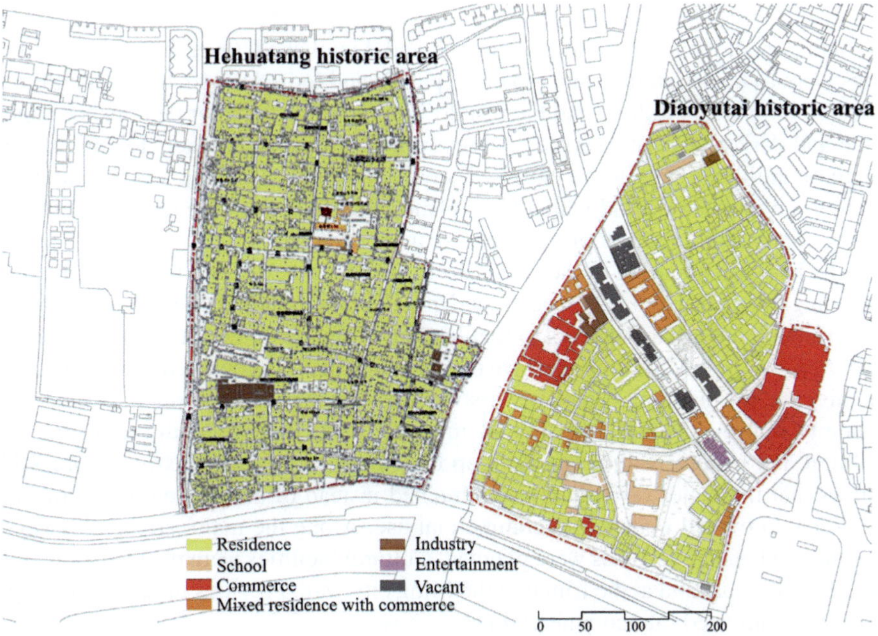

Fig. 7.8 Fifth attribute: building function

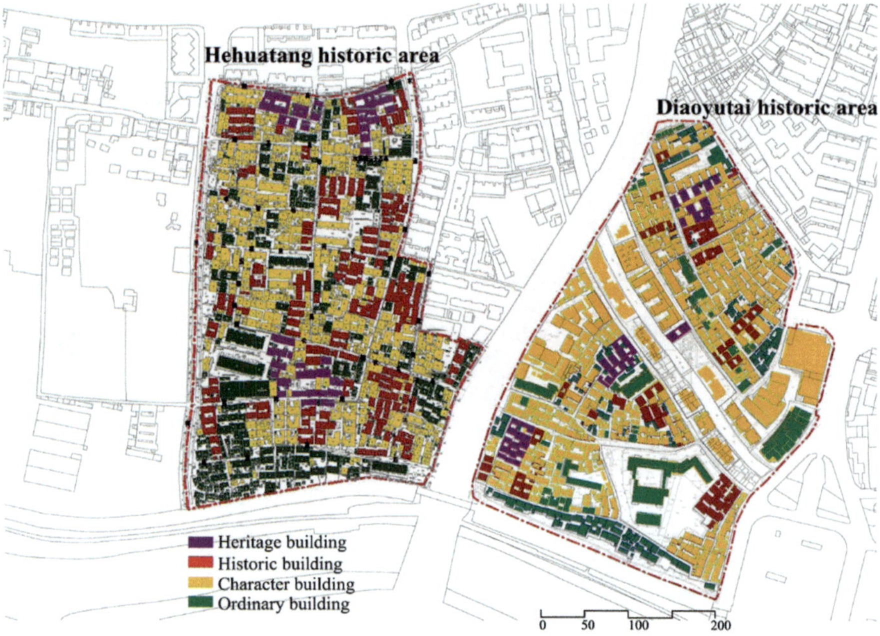

Fig. 7.9 Evaluations of building values

be maintained, repaired, renovated, altered and demolished. Over 40% of the buildings were to be maintained and repaired (purple and red). Around 40% of buildings shall be renovated (yellow) and the rest 20% were to be altered or demolished (light blue and dark blue). This served as a reference for decision-makers to control and regulate buildings within the enclosed area, especially regarding which buildings were to be maintained and which to be changed to cope with the tensions between rising contemporary development pressure and conservation.

The land use planning of the two historic areas was developed from the Detailed Planning of Old Chengnan District issued in 2012. This is a fundamental tool for local government and urban planners to regulate land use and guide future development, according to the city's comprehensive planning and district's detailed planning goals. In this case, the planning goals were firstly to maintain the area's residential function, while to improve its urban vitality by injecting more uses; secondly to display its cultural and historic values and develop the cultural tourism industry. Figure 7.11 shows that the major residential use remained instead of the conventional practice of doing an overall update into commercial use. Along the major street alleyways, commercial and mixed uses were planned. Cultural facilities were planned along the River Neiqinhuai and at the fringe of the Ming city wall. Previous industry buildings were to be updated into commercial use.

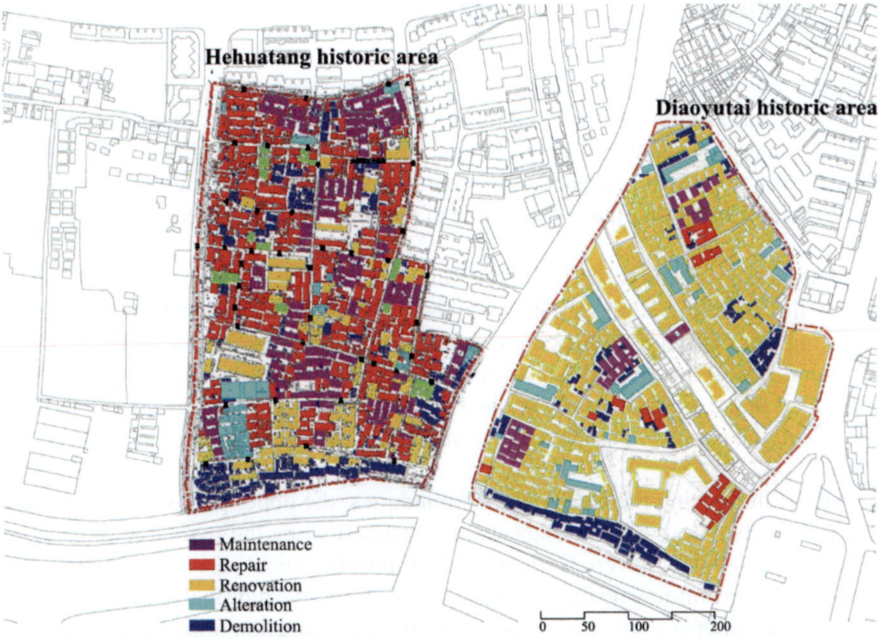

Fig. 7.10 Physical interventions to buildings

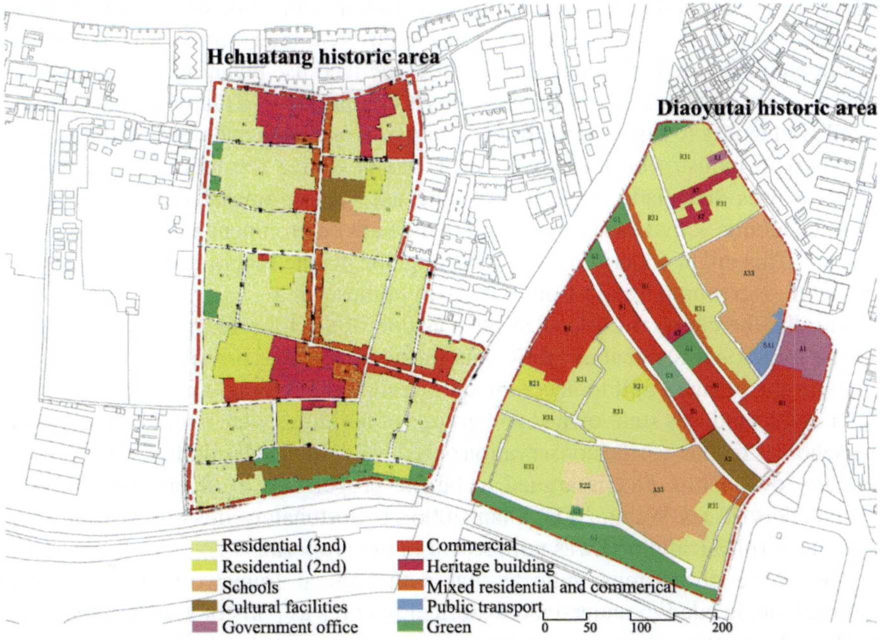

Fig. 7.11 Functional planning of both areas

7.3 Discussion

This chapter has provided an example of the complexities and tensions that exist between conservation and redevelopment through one historic district in Nanjing, China. It describes the evolving change from 'massive demolition and massive construction' to 'small-scale and incremental urban redevelopment' since the 2010s. The local conservation planning practice of two historic urban areas is introduced in detail.

Many scholars (Yao & Zhu, 2009; Zhang & Zhou, 2001; Zhang et al., 2011) have argued that there is a study gap between the historic district and historic urban area, as the historic district usually occupies several km^2 while the size of the historic area is around 20–30 ha. How the historic urban areas are physically and functionally embedded in the city is not fully studied. This weakness has influenced the rigour in delimiting the conservation area and establishing the management priorities, as put below.

Firstly, in drawing the conservation boundaries, the upper part of Diaoyutai area which was also the traditional fabric was excluded while the two large plots along Zhonghua Road which were formed in other morphological periods were included.

Secondly, in building up the management priorities, the physical intervention hierarchies of maintenance, repair, renovation, alteration, and demolition (Fig. 7.10) and the functional planning of commercial, school, cultural facilities and green (Fig. 7.11) were developed separately, rather than combined together. There was a mismatch between the physical form and the planned function in some areas. For example, one area was planned to be a school, yet the physical intervention was to renovate traditional residential buildings. Meanwhile, the land use planning was a top-down approach, based on the objective of the city's comprehensive planning and district's detailed planning to improve the living quality on one hand and develop the cultural and tourism industry on the other. Yet it was at coarse resolution, and separated different uses and made large areas single use. This proposal ignored both how socio-economic activities are organized in the built environment and what the potentials are. This might undermine the viability of the planned socio-economic activities or lead to overcommercialization or tourism in these areas.

Thirdly, the courtyard (physical plot) as one of the form elements was missing. As the traditional residential fabric was formed under the private land ownership, plot was the planned unit to accommodate groups of buildings for each household. After 1949, land was state-owned and groups of buildings were assigned to different households. To manage the conservation or redevelopment of buildings with complex property ownership, plot is argued to be the proper unit to incorporate different stakeholders (Lin & Wang, 2003; Shan, 2006). Unfortunately, this information about courtyards (physical plots) was not present in the proposal.

The above discussions indicate that a theoretical framework and analytical approach are needed to understand historical features as parts of a wider landscape, and how the existing socio-economic activities are organized in the built environment and what the potentials are in preparing the conservation proposal. The next chapter will discuss how the integrated morphological framework could provide reference for urban landscape management, in particular delimiting the conservation area and building up the management priority.

References

Bandarin, F., & Van Oers, R. (2012). *The historic urban landscape: Managing heritage in an urban century*. Wiley.

Chen, F. (2011). Traditional architectural forms in market oriented Chinese cities: Place for localities or symbol of culture? *Habitat International, 35*(2), 410–418.

Fang, Z., Hua, X., Liu, Q., Xu, Y., Du, L., & Chen, W. (2012). Jiucheng baohu de zaisikao (Reflections on the conservation of the historic city of Nanjing). *Shijie Jianzhu Daobao, 2*, 10–33. (in Chinese).

Han, P. (2004). *Qinhuaihe shihua (History of Qinhuai river)*. Nanjing chubanshe (in Chinese).

Jacobs, J. (1961). *The death and life of great American Cities*. Wiley.

Koh, M., & Melic, K. (2017). *A historic heart how heritage districts can make cities more liveable*. Retrieved from Singapore: https://www.clc.gov.sg/docs/default-source/reports/a-historic-heart.pdf.

Lin, N., & Wang, K. (2003). Wenhua chuancheng yu chengshi fanzhan (Urban development and cultural identity—A case study of Beijing Nanchizi historic urban area). *Jianzhu xuebao* (11), 7–11 (in Chinese).

Liu, J., & Shen, Y. (2010). Zouxiang lixing de nanjing lishididuan baohu guihua (The evidenced-based Nanjing conservation planning of historic and cultural area). *Xiandai Chengshi Yanjiu, 12*, 50–54. (in Chinese).

MOHURD. (2018). *Guanyu zai shishi chengshi gengxin xingdongzhong fangzhi dachaidajian wenti de tongzhi (Announcement on preventing mass-demolition and mass-construction in the urban redevelopment practice)* (in Chinese).

Ruan, Y. (2004). Duiyu woguo lishi jiequ baohu shijian moshi de pouxi (The study on the conservation managment of historic urban areas). *Tongji Daxue Xuebao, 15*(5), 1–6. (in Chinese).

Ruan, Y., & Sun, M. (2001). Woguo lishijiequ baohu yu guihua de ruogan wenti yanjiu (Studies on the conservation planning of Chinese historic urban areas). *City Planning Review, 25*(10), 25–32. (in Chinese).

Shan, J. (2006). Cong daguimo jiufang gaizao dao xunxi jianjin youji gengxin (From mass demolish to incremental redevelopment). *Wenwu* (7), 26–40 (in Chinese).

Shen, J., Zheng, X., & Wang, Q. (2013). Laocheng konggui daodi kongshenme (Explanation on Nanjing's detailed planning of the old city). *Chengshi Guihua, 9*, 88–93. (in Chinese).

Wang, L., & Mo, C. (2017). Chengshi gengxin he fengmao baohu de chengshi sheji yu chengshi zhili shijian (Urban design and governance for urban renewal and landscape preservation). *Guihuashi, 33*(10), 135–141. (in Chinese).

Whitehand, J. W. R. (2007). *Conzenian urban morphology and urban landscapes*. In *Paper presented at the Proceedings, 6th International Space Syntax Symposium*, Istanbul.

Wu, F. (2002). Real estate development and the transformation of urban space in China's transitional economy, with special reference to Shanghai. In *The new Chinese city: Globalization and market reform* (pp. 153–166).

Wu, L. (1994). *Beijing Jiucheng yu jier hutong (The old city of Beijing and Juer Hutong)*. Zhongguo jianzhu gongye chubanshe.

Yao, D., & Zhu, G. (2009). Lishi wenhua jiequ baohu guihua de shifanxing yu shiyingxing yanjiu (Study of the conservation planning of historic and cultural area). *Jianzhu Xuebao* (6), 40–43 (in Chinese).

Zhang, J. (2012). Lishi diduan jianzhu zonghe pingjia yanjiu (The evaluation of historic urban area). *Shanxi Jianzhu, 38*(15), 8–9. (in Chinese).

Zhang, K., & Zhou, J. (2001). Faguo chengshiguihua bianzhi tixi dui woguo de qishi (Learning from the planning propsals of Paris's historic areas). *Chengshi guihua* (8), 37–41 (in Chinese).

Zhang, Y., Liu, Q., & Zhang, J. (2011). Lishi chengqu chiduxia de lishididuan xitong baohu zhengzhi gongzuofangfa tantao (The study of the conservation strategies of historic urban areas-based on the Conservation Planning of Old Chengnan). *Chengshi guihua lunwenji*, 8295–8305 (in Chinese).

Zhong, X., & Chen, X. (2017). Demolition, rehabilitation, and conservation: Heritage in Shanghai's urban regeneration, 1990–2015. *Journal of Architecture and Urbanism, 41*(2), 82–91.

Zhou, L. (2011). Positive protection and integrate innovation of Chinese Historic and CulturalCity. *China Acient City* (3), 4–12 (in Chinese).

Zhu, L., & Goethert, R. (2010). Different approaches in conservation of historic cities in China. *Proceedings of the Institution of Civil Engineers, 163*(3), 189–196.

Chapter 8
An Integrated Morphological Approach to Urban Conservation and Management: The Case of Nanjing

Abstract Moving onto to conservation practice, combining the historico-geographical and the configurational approaches can generate a constructive negotiation between preserving historicity and enhancing sustainable use of urban spaces. The analytical method that combines the two morphological approaches has been applied to two historic areas—Hehuatang and Diaoyutai—in Old Chengnan district. It further elaborates how the evaluation of landscape units in terms of historicity and spatial potential could inform urban landscape management, especially the sharper identification of conservation priorities. A comparison will be made between the management proposal developed by the local planning authority and that suggested by the integrated approach. It is found that the integrated approach has the potential to provide sound reference for urban planners and designers to manage change in a small-scale and incremental way and contribute to a synergetic view between conservation and urban redevelopment.

Keyword Landscape unit · Historicity · Spatial potential · Conservation and management priority · Small-scale urban conservation and redevelopment

8.1 Evaluation of Landscape Units

This chapter, based on Chap. 6 examination of the spatial potential of landscape units, will explain how this evaluation could provide reference for urban planners to manage change, especially what new forms to be incorporated in the old urban landscape, and what locational and functional choices to be suggested. The way in which the analytical framework that could strengthen the theoretical and methodological foundation of area-based conservation in Nanjing, will also be demonstrated by comparing it with the conventional conservation plan in Chap. 7.

As explained in Sect. 6.3.1, in the landscape units of the market-oriented economy period (1998–2015), Hehuatang and Diaoyutai historic areas belong to the first-order landscape unit of traditional settlement (II). There are a few adjacent large plots and building blocks developed in different periods, such as Ia and Id (Inner fringe of River Neiqinhuai dating back to the Feudal period), IVm (linear commercial development in the 1990s) and Vq (linear commercial development in the 2000s). Their

X. Li and Y. Zhang, *Conserving and Managing Historical Urban Landscape*, The Urban Book Series, https://doi.org/10.1007/978-981-19-4222-8_8

compositions of plots, buildings and streets are different from that in the traditional settlement. Thus, they are excluded from the first-order landscape unit of traditional settlement (II).

The second-order landscape units of traditional settlement (II) are composed of Mendong (IIe) and Menxi (IIf), which are divided by the north–south axis Zhonghua Road. The historic areas of Hehuatang and Diaoyutai are in the third order, which are IIf22 and IIf23. When zooming into these two areas, it becomes clear that the majority of these urban areas is composed of traditional courtyard buildings, street alleyways and narrow-frontage plots. All these three form elements help define a unique streetscape, indicating local distinctiveness and sense of place in this area (Fig. 8.1). The rest of these urban areas are composed of several multi-storey housing blocks built after 1979. Their plot size is larger, and the large building blocks are in contrast with the one-or-two-storey traditional houses (Fig. 8.2). The traditional residence (IIf22①) and multi-storey housing (IIf22②③…) together compose the 4th tier landscape units of traditional settlement.

The historico-geographical analysis has produced a four-tier division of urban landscape, providing an enhanced understanding of the city as an interrelated entity and of how different component parts related to each other and form the whole settlement (Fig. 8.3). This provides a sound basis for drawing the boundaries of Hehuatang and Diaoyutai historic areas. It also provides morphological guidelines for future physical interventions in each landscape unit. For example, the multi-storey housing

Fig. 8.1 Street views of Yinma street alleyway, Dabaihuaxiang alleyway and Hehuatang alleyway (photos taken by authors)

Fig. 8.2 Street view of multi-storey housing in IIf22 (photos taken by authors)

unit consists of large-size plots and building blocks, which is incompatible with its surrounding traditional courtyard housing unit. In the future, sensitive adaptions of this multi-storey housing unit are suggested in order to maintain the integrity of the historical urban landscape.

Meanwhile, as discussed in Sect. 6.3.2, the spatial potential measured by the configurational approach could articulate how socio-economic activities are organized in the built environment and what the potentials are. Spatial potential is defined by two spatial metrics: spatial accessibility and spatial capacity. These two metrics were summarized at the unit of physical plot and aggregated at the 1st, 2nd, and 3rd tier landscape units. The combination of spatial accessibility and spatial capacity has identified the varied spatial potentials of landscape units: matched, in-between and mismatched.

In the 1st tier landscape unit of traditional settlement (II), the major function is residence, yet its neighbouring work and visit functions could contribute to its functional balance (77:13:10). Its spatial accessibility is low compared to other landscape units (1.022), because traditional settlement is segregated in the superimposed

Fig. 8.3 Landscape unit division of Hehuangtang and Diaoyutai historic areas (IIf22 and IIf23)

large urban grids. Thus, its spatial potential is mismatched, indicating its low spatial accessibility may not fully support its high functional potentials.

In the 2nd tier landscape units of Menxi (IIf), it has mixed use (74:15:11) because of its close proximity to the commercial developments along Zhonghua Road. The rich work and visit functions nearby contribute to its balanced functional potentials. Yet its spatial accessibility (1.039) is low. Its spatial potential is also mismatched.

In the 3rd tier landscape units of Hehuatang (IIf22), it has bi-functional uses (83:11:6) while Diaoyutai (IIf23) has mixed use (62:19:19) because Diaoyutai is closer to the linear commercial development along Zhonghua Road and Zhongshan Road. The spatial accessibility of both areas is low. The spatial potential of Hehuatang is in-between, and that of Diaoyutai is mismatched.

In the 4th tier landscape units of traditional residence in Hehuatang (IIf22①), there are clusters of physical plots with varied spatial potentials, subdividing the lowest hierarchy of landscape units, as shown in Fig. 8.4.

For clusters with mismatched spatial potential, there are two types: the first (green dotted pattern) is high spatial accessibility but low-spatial capacity. It is mainly distributed along Xiegongci-Yinmaxiang and Shuizhaian alleyways, which are the main alleyways in Hehuatang area. The second (green) is high spatial capacity but low spatial accessibility. It is deeply embedded in Diaoyutai area and has rich work and visit functions nearby.

For clusters with in-between spatial potential, there are also two types: the first (blue) is higher spatial accessibility and lower spatial capacity; the second (purple) is lower spatial accessibility and higher spatial capacity. The first type within IIf22① has medium spatial accessibility and low spatial capacity, while within IIf23①, there is high accessibility and medium spatial capacity. The second type within IIf22① has low accessibility and medium spatial capacity, while within IIf23①, there is medium accessibility and high spatial capacity. These two types constitute the majority of the historic areas.

For clusters with matched spatial potential, there are three types: the first (orange grid pattern) has both high accessibility and high spatial capacity; the second (orange) has both medium accessibility and medium spatial capacity; the third has both low accessibility and low spatial capacity. The first is mainly located in IIf23①, along Diaoyutai, Tangfanglang and Xuanzixiang alleyways. They have commercial developments nearby and the spatial accessibility ensures enough movement flows to support such functional diversities. The second is in both IIf22① and IIf23①, along Yingaoxiang, Mingyangli and Liujiaojing alleyways, where offices and schools are located nearby. The third is in IIf22①, which is located at a distance from the main alleyways, and has mono-residential use. The low spatial accessibility can restrict and regulate movement flows, allowing a relatively quiet and private environment for local residents.

By combining both historicity and spatial potential, the strengths and weaknesses of landscape units can be evaluated, and this evaluation could provide reference for future management of change, such as how to harmonize the ordinary or character buildings with their surroundings to ensure the integrity of urban landscape, and

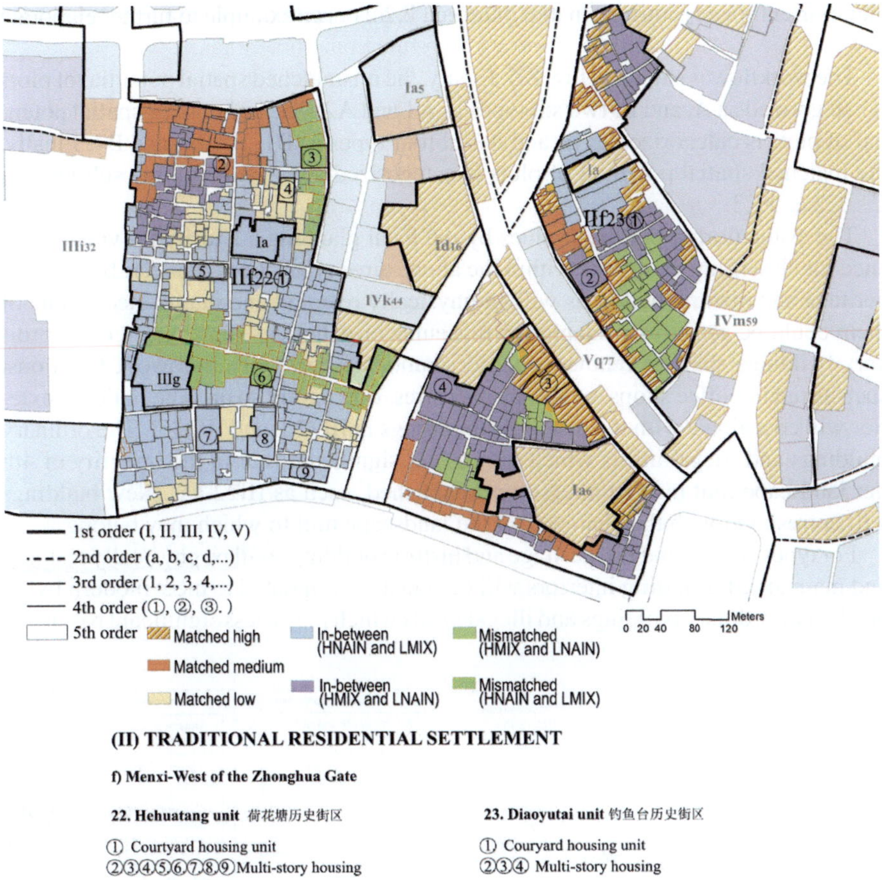

Fig. 8.4 Classification of various spatial potentials of urban areas within the 4th tier landscape units (IIf22 and IIf23)

how to maximize the physical potential to accommodate socio-economic activities to improve people's lives.

8.2 Development of Management Proposal

How this analytical framework can be applied to help preparing a management proposal is elaborated in this section by using the two historic areas (IIf22① and IIf23①) as an example. Though the current analytical unit is at the scale of physical plot, this integrated framework has the potential to further zoom in on the building scale. This section adopts the building evaluation framework developed by the local planning authority (outlined in detail in Sect. 7.2.2) as an example to further elaborate it.

The workflow is shown in Fig. 8.5. Firstly, the mismatched spatial potential of plots is categorized as A, and its two subtypes as A1 and A2. The in-between spatial potential of plots is categorized as B, and its subfour types as B1, B2, B3 and B4. Finally, the matched spatial potential of plots is categorized as C, and its three subtypes as C1, C2 and C3.

The evaluation of building values by the local planning authority provides reference for classifying buildings within the above three types. For types A1, B1 and B2, heritage and historic buildings will be physically restored and maintained. Some of them could be sensitively adapted to museums or galleries. The character and ordinary buildings shall be renovated to accommodate more visit and work functions, such as retail, coffee shops, community centres, offices and so on. The building exterior will coordinate with the rest of the buildings in IIf22① and IIf23①. The ordinary buildings and illegal infills which play a less significant role in the integrity of 4th tier landscape unit may be altered or demolished, such as IIf22③⑥. New buildings will correspond with the character of the landscape unit to which they belong.

For types A2, B3 and B4, heritage and historic buildings will be physically restored and maintained. Building interiors will be sensitively updated to meet modern living needs. The ordinary buildings and illegal infills which play a less significant role in the

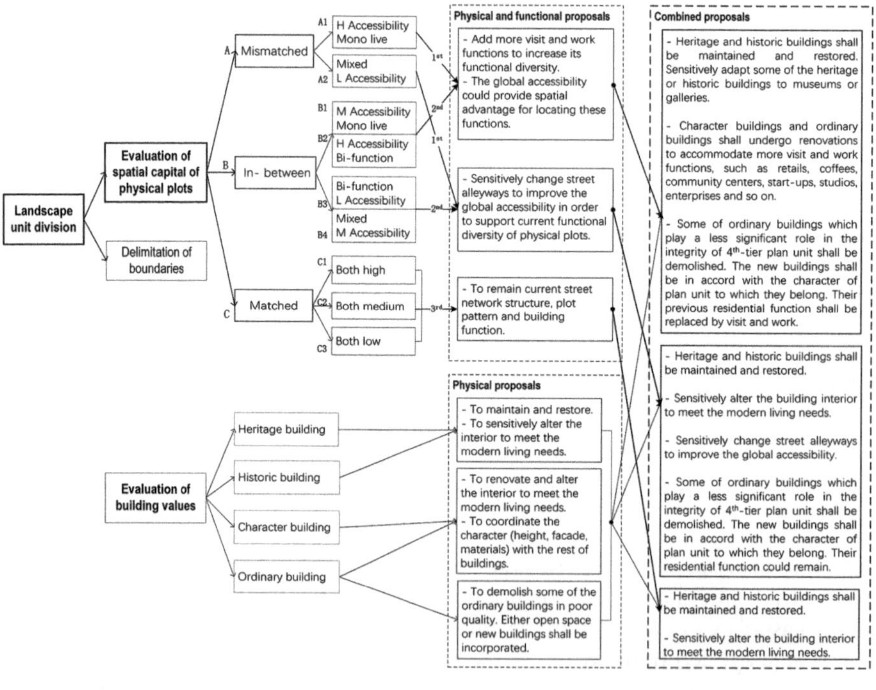

Fig. 8.5 Workflow of the management of change

integrity of 4th tier landscape units could be altered. New buildings will correspond with the character of the landscape unit to which they belong. Their residential function could remain. When it comes to street networks, it has been regulated that historical streets and alleyways shall not be altered or widened in the *Guidelines on Conservation Planning of Precious Historic and Cultural City of Nanjing*. Yet for ordinary streets and alleyways, sensitive changes could be made to improve the accessibility to areas of A2, B3 and B4. New street segments (highlighted in blue colour) are proposed to add in order to connect the previous dead-end alleyways by cautiously removing existing ordinary buildings and illegal infills or widening existing passageways (See Fig. 8.7).

For types C1, C2 and C3, heritage and historic buildings will be physically restored and maintained with building interiors sensitively updated to meet modern living needs. The street alleyways and current function could remain unchanged for the time being.

Based on the evaluation of both the historicity and the spatial potential, the lowest hierarchies of landscape units have been disaggregated into several clusters with distinctive intervention strategies (Fig. 8.6).

8.3 Comparison and Discussion

The evaluation of the spatial potential that combines the historico-geographical and the configurational approaches in this study has addressed the three deficiencies of conservation planning by the local planning authority put forward in Sect. 7.2.2. Firstly, with regard to the conservation boundary, most of the 1st tier boundaries delineated by the historico-geographical approach correspond closely to the boundaries delineated by the planning authority, but the former also includes the upper Diaoyutai area, which was ignored in the planning authority's proposal. While Ia, Id, IIIg, IVm and Vq are excluded in this study, because they represent the morphological character in different morphological periods. The 3rd tier boundaries define Hehuatang (IIf22) and Diaoyutai (IIf23) as a whole, as they belonged to one historic area previously. This indicates that the interventions might need to strengthen their in-between connections. The 4th tier boundaries distinguish modern multi-storey housing from traditional courtyard housing within Hehuatang (IIf22) and Diaoyutai (IIf23) areas. The hierarchical division could uncover how different parts relate to one another and provide evidence for physical interventions at different levels of landscape units before zooming in to the building level.

Secondly, regarding the socio-economic performance, the analytical framework could provide evidence for incremental change by measuring the spatial potential of landscape units. Clusters of urban areas with varied spatial potentials are identified within the 4th tier landscape unit—matched, in-between and mismatched—providing reference for small-scale and incremental future interventions, either improving the spatial accessibility to support the functional diversity or injecting new uses to take full advantage of spatial location. However, in the planning authority's management

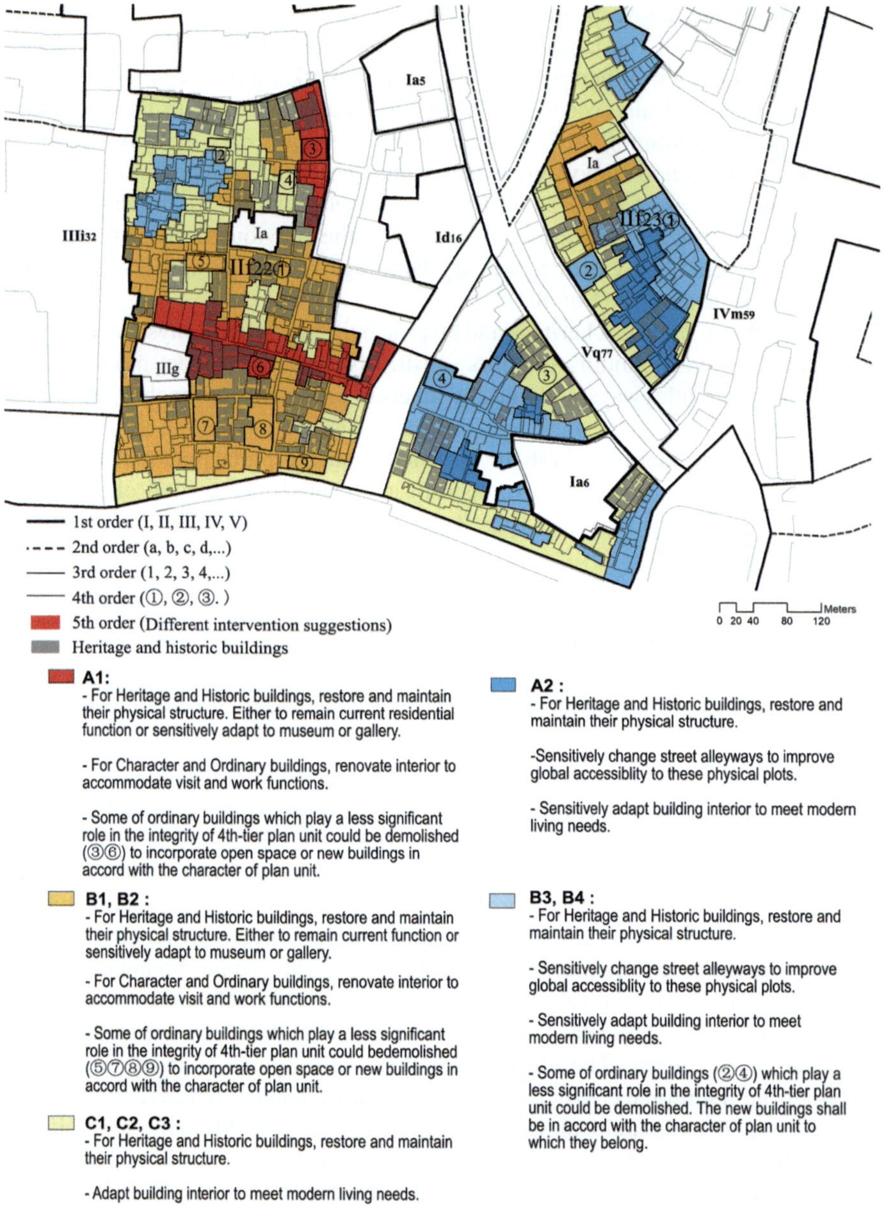

Fig. 8.6 Conservation proposal based on the integrated historico-geographical approach and configurational approach

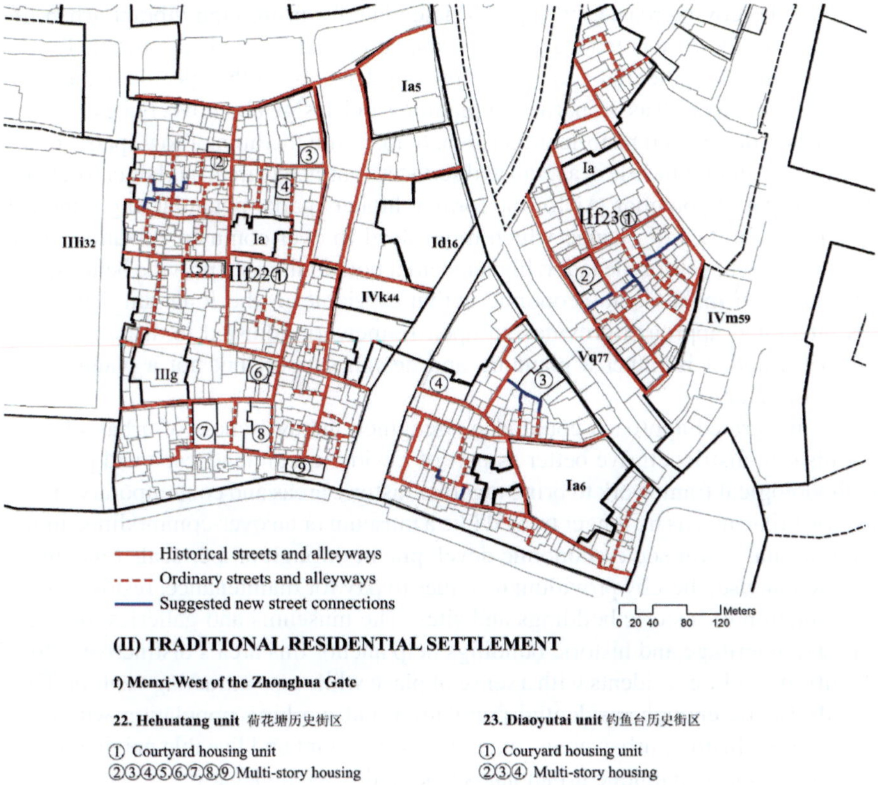

Fig. 8.7 Suggested new street connections within areas of A2, B3 and B4

proposal, the top-down land use planning is at a coarse resolution and ignores the potentials of urban form to accommodate socio-economic activities (Fig. 7.11). For example, the Xiaoshunli street alleyway was planned to accommodate mixed uses on both sides in the planning authority's proposal, yet it has only medium accessibility. It may lack spatial advantage for these mixed uses to survive and prosper in a long run.

Thirdly, physical plot (courtyard) features as both the analytical unit and the management unit. Researchers argue that physical plot is the proper unit to incorporate different stakeholders to manage the conservation or redevelopment of buildings under complex property ownership. This unit might enhance the practicability of management proposal. Yet in the planning authority's proposal, the management priorities are often building-based, and there lacks description or assessment of the historicity or socio-economic performance at the scale of physical plots (courtyards).

To sum up, compared to the conventional planning practice, the evaluation of spatial potential of landscape units could provide sound reference for urban planners and designers to manage change in a small-scale and incremental way. For instance, which interventions could improve the spatial accessibility or spatial capacity, thereby increasing the spatial potentials for promoting vibrant urban life. However, it should be noted that higher spatial potential is not always better. To what extent the spatial potential could be improved depends on the specific context and circumstances as well as the particular planning and design objective. For example, in a residence-dominated historic urban area, to maintain medium or low spatial potential might be enough while in a retail and service-dominated area, it is better to achieve matched spatial potential. As for the former, higher spatial potential might indicate higher possibilities of change which might lead to over-commercialization in this residential area, yet for the latter, higher spatial potential might provide better spatial advantages to make socio-economic activities viable in the long run. Therefore, this integrated approach provides an open framework adjusted to different design scenarios, and it is rather a planning and design support but not a planning and design director.

The integrated approach could also strengthen the awareness of urban planners that historic districts thrive better as part of a wider urban landscape and provide a methodological framework to bring together historic areas and contemporary uses. A historic urban area is no longer treated like a museum or an over-commodified image, but as a catalyst for socio-economic development through proper-scale tourism and commercial use, thereby providing revenues to pay for maintenance, restoration and rehabilitation of historic buildings and sites. The museums and galleries located in conserved heritage and historic buildings help anchor this area's distinctive cultural identity, providing residents with a sense of place while also attracting visitors. There would also be enough residential population and working population within and around these historic urban areas, maintaining a vibrant and liveable neighbourhood 24/7 and ensuring a unique urban landscape as the lasting legacy.

Chapter 9
Conclusion: The First Step in a Long Journey

Abstract This book has evaluated and corroborated the theoretical groundwork of combining the historico-geographical and configurational approaches through empirical studies of Ludlow (UK), Chinatown (Singapore) and Nanjing (China). It has explored how this framework could provide reference for urban planners and designers to manage change in historical urban landscape. It is just a first step in a long journey towards combining and coordinating different morphological approaches to enhance the understanding, explanation and prescription of urban form. More rigorous and more empirical-based studies could be carried out to verify and enrich this analytical framework in the future.

Keywords Urban morphology · The historico-geographical approach · The configurational approach · Fast-changing urban landscape

This book started by introducing the current challenges facing urban morphology, that is, combining and coordinating different morphological traditions to enhance the understanding, explanation and prescription of urban form. While the disciplinary and language differences among the different morphological traditions create considerable hurdles, the common ground shared by these traditions does provide a concrete basis for combining and coordinating them, such as Moudon's (1997) summary of three fundamental components (form, resolution and time) and Kropf's (2009) summary of form elements and patterns. There have been extensive explorations of combining the historico-geographical and the process-typological approach due to their similar historical roots (Maffei & Whitehand, 2001). In recent years, there have been increasing attempts to coordinate different morphological approaches (Kropf, 2014; Monteiro & Pinho, 2021; Oliveira, 2015; Scheer, 2016), especially the historico-geographical and the configurational approach (Griffiths et al., 2010; Pinho & Oliveira, 2009). The theoretical explorations of combining these two approaches are relatively rich, yet empirical studies are generally rare. It still remains unclear in which way and to what extent these two approaches could be combined, and this integrated approach requires empirical substantiation in different urban contexts.

X. Li and Y. Zhang, *Conserving and Managing Historical Urban Landscape*, The Urban Book Series, https://doi.org/10.1007/978-981-19-4222-8_9

Meanwhile, the rapid growth of cities in the past fifty years have created great opportunities to enhance the quality of life, access to services and technology benefits, yet it has also created problems (Bandarin & Van Oers, 2012). One of them is the fast-changing urban landscape, where large-scale and mono-lithic development projects take place, demolishing the traditional fabric as well as the underlying intricate functional and social microstructure in the historic inner cities around the world (Chen & Thwaites, 2013). The historic character of cities has been greatly threatened. How to resolve the conflicts between the preservation of urban heritage and socio-economic activities has become a major issue in urban planning and urban governance. However, researchers have long argued that historic districts not only have historic and cultural values that should be preserved, but they also could serve as a powerful catalyst for economic and social regeneration (Bandarin & Van Oers, 2012; Jacobs, 1961; Koh & Melic, 2017). Conservation of historic districts extends from their heritage values to a broader context of urban management and development (Bandarin & Van Oers, 2012; Fang et al., 2012; Zhang et al., 2011). Through careful small-scale, incremental modifications of form and function as a source for changes, the development of diverse economic activities and an improvement in the quality of life could be achieved (Fang et al., 2012; Scheer, 2010; Shan, 2006; Wang & Mo, 2017; Wu, 1994; Zhang et al., 2011).

Yet in the planning practice, historic features tend to be treated as isolated elements. There is little sense of how they fit together to form an interrelated entity and are part of a process of change (Whitehand et al., 2011). One of the key consequences is that ill-informed geographical boundaries of historic area in management proposals usually cut across landscape units, which are products of a city's historical development (Whitehand, 2007). Another key consequence is that management priorities are largely focused on individual buildings, ignoring the relation between historic buildings and historic urban area they are part of, which often results in a fragmented urban landscape. Moreover, land use zoning is still perceived as a prevalent and efficient tool for urban planners to plan new or renewal projects. In reality, it separates different uses and prevents a diverse urban community. Massive areas created are devoted to single uses (Christopher, 1966; Scheer, 2010). There lacks a sound theoretical basis and methodology to explore the historicity and socio-economic performance of urban form and to manage urban landscape with regard to what to conserve and what to change. However, morphological approach has demonstrated its power in the recognition, delimitation and management of historical urban landscape within its contemporary context (Garyfalia, 2017; Whitehand, 2015).

The historico-geographical approach could articulate the historical development of an area through a systematic mapping analysis of ground plan, building fabric and land use. There have been applications, adaptations and demonstrations of this approach in both landscape management and conservation planning in the UK and in China, especially in delimiting the boundary of historic area and identifying conservation priorities (Baker, 2010; Gu, 2010; Kropf, 1996, 2011; Larkham, 1996; Thomas, 2018; Whitehand & Gu, 2007a; Whitehand et al., 2011; Whitehand & Gu, 2007b). Yet the historico-geographical approach is not enough, as it views 'historicity' as the key factor of urban landscape, which outweighs its contemporary value.

The configurational approach could identify lively local centres which are socially and economically viable based on the relationship between spatial configuration and urban movement (Yang & Hillier, 2007). It could also analyze how socio-economic activities are organized in the built environment, and what the potentials are by linking spatial configuration with other socio-economic data (Bobkova et al., 2017; Marcus, 2007, 2010; Marcus et al., 2017). It could guide future interventions to improve socio-economic performance and reduce the physical disruptions to historic quarters caused by the urban regeneration process (Karimi et al., 2007; Stonor, 2017).

Combining the historico-geographical and configurational approaches could have the potential to contribute to a balanced understanding of both the historical character and the socio-economic potentials of urban areas and to management of fast-changing landscape in a small-scale and incremental way. Based on such discussions, this book evaluated and corroborated the theoretical groundwork of combining the historico-geographical and configurational approaches through empirical studies of Ludlow (UK), Chinatown (Singapore) and Nanjing (China).

An analytical method that combines and coordinates the historico-geographical and the configurational approach was developed to provide an enhanced understanding of the historic development and the socio-economic performance of cities. Based on the enriched understanding, this book presented an evaluation of the strengths and weaknesses of urban landscape and explored how this framework could provide reference for urban planners and designers to manage change in historical urban landscape, in particular what new urban forms to be incorporated, and what the functional and locational choices of new development are suggested.

9.1 Summary of Key Findings

The key findings of this study are twofold:

Firstly, the analytical framework that combines the historico-geographical approach and the configurational approach could contribute to a clearer and more accurate plan unit recognition. On one hand, the configurational approach could uncover the generic generative rules of cities which could explain urban growth and development to some extent. For example, it was found out that the higher hierarchies of plan units in different morphological periods tend to form along the globally most integrated street segments. As shown in the case studies of Ludlow, Singapore and Nanjing, the 1st and 2nd tier plan units tend to form along the 20% of globally most integrated street segments, which are the long and continuous lines constituting the foreground network of the city. This suggests that although the three cities were developed according to different planning decisions in different contexts, their urban layouts exhibit a similar development process, that is, urban growth tends to take place along the most accessible streets to take advantage of urban movement to foster developments. The slight difference is that the local integration measure in Ludlow could correspond to more plan units of the lower tier than that in Singapore and Nanjing. As Ludlow grows incrementally outward from the existing periphery of the mediaeval town, urban settlement in each morphological period exhibits different street layouts—cul-de-sacs and curvilinear street segments. Thus, street configuration alone could capture lower-tier plan units. While Nanjing and Singapore have undergone intensive urban redevelopments, large urban blocks in gridiron street layout emerge in the urban areas of Old Chengnan and Chinatown separately. As a result, the local urban settlement in each morphological period could be more clearly characterized by the compositions of plots and building blocks than the generic planned urban grid. Therefore, street configuration plays a relatively limited role in lower-tier plan unit recognition. From the above discussion, it can be argued that besides cartographic sources, plot-by-plot survey and documentary evidence, the configurational approach has the potential to provide an additional reference to plan unit recognition in an objective manner, and its usefulness is dependent on the specific street network patterns.

On the other hand, the historico-geographical approach could provide place-specific contexts for each city to help explain the changes in the configurational values of different plan units to a certain extent. As the global NACH values of plan units indicate how integrated and structured they are with the other parts of city, the historico-geographical approach could address the difference by elaborating on the development layouts of Ludlow, Chinatown and Nanjing. For example, in Ludlow, the newly developed plan units outside the Mediaeval town tend to have lower global NACH values than plan units within the Mediaeval town. The underlying reason is that the new development was located in areas adjacent to the old organic structures out of the town, and it is less integrated with the other parts of the city than the plan units in the historic core of the town. Yet in Nanjing and Singapore, it is not always the case: the newly developed plan units tend to have higher or lower global NACH values than the traditional plan units. For the former condition, the new development was always imposed on the old fabric, which was well embedded within the city; therefore, it is more globally integrated. For the latter, the new development was the outward urban growth and expansion; thus, it is less globally integrated. In the local scale, traditional urban fabric in these three cities tend to have higher integration values due to its fine grain and high-street density, such as high-street unit in Ludlow, Kreta Ayer area in Chinatown and Hehuatang area in Nanjing. Yet not all local settlements can be recognized by the configurational patterns or values. Other factors, such as development strategies and planning regulations and guidelines, may outweigh the impact of street configuration in the transformation of plan units.

It can therefore be seen that the combination of the historico-geographical and configurational approaches is capable of providing enhanced understandings of the historical transformations of urban form. It is conditioned by the external social-economic factors; at the same time, it is also underscored by the more generic generative rules of urban space. The extent to which these two aspects influence the transformation of urban form is dependent on the different development process of urban areas—incremental growth or fast-changing urban landscapes.

Secondly, the analytical framework could provide a cross-scale evaluation of the strengths and weaknesses of landscape units in terms of historicity and spatial potential. Spatial potential in this study refers to how urban form could support, impede or organize socio-economic activities. It consists of two spatial metrics – spatial accessibility and spatial capacity. Spatial accessibility is measured by the configurational approach, and it reveals the locational potentials of urban form for supporting or impeding diverse socio-economic activities. Spatial capacity is measured by linking the configurational analysis with socio-economic data (emerging open data POI, land use data), which reveals the ability of urban form to accommodate socio-economic activities within a certain radius. By combining both spatial accessibility and spatial capacity, it contributes to a classification of spatial potentials into matched (both high, medium or low spatial accessibility and spatial capacity), in-between (high spatial accessibility and medium spatial capacity, or medium spatial accessibility and low spatial capacity or vice versa) and mismatched (high spatial accessibility but low spatial capacity or vice versa).

Based on the overlapped mappings of landscape unit and spatial potential, the lower hierarchies of landscape units were further divided into clusters of areas with the same spatial potential compared to conventional land use data. Meanwhile, an analytical framework for evaluating the strengths and weaknesses of landscape units was developed based on both historicity and spatial potential. This evaluation is twofold: on one hand, it examines what is important to the historic character and whether existing urban forms accord with the character of the urban landscape units to which they belong; on the other hand, it identifies spatial metrics with low values, for example, poor spatial locations which may not support diverse content or areas with mono-function which may not fully take advantage of their locational advantage. Future interventions are proposed accordingly to improve this area's vitality.

To what extent the spatial potential could be improved depends on the specific context and policy as well as the particular planning and design objective. For example, the conservation guidance speaks of 'positively managing the development of historic urban areas and enhancing socio-economic sustainability'. For areas with mono-residential function, new functions shall be sensitively selected and injected so as not to be in conflict with existing residential use. For former industrial areas, government and local communities shall play a role in regulating and overseeing the market force to prevent over-commercialization or gentrification in these areas in the long run. For mixed-use areas with poor spatial accessibility, new street segments could be cautiously added to make dead-end alleyways connect in order to attract more people to these areas and support current functional diversity. From this perspective, this integrated approach provides an open framework adjusted to different design scenarios.

9.2 Contributions

The contributions of this study are threefold:

Firstly, the empirical studies of Ludlow, Singapore and Nanjing have verified and corroborated the theoretical groundwork of combining the historico-geographical and the configurational approach, suggested by Kropf (2009) and Griffith (2010). They both assumed that the combination of the two approaches could uncover the generative process of urban formation and transformation, as both the historic character and socio-economic character of urban areas are rooted to an extent in the structure of urban form. Building on their theoretical assumptions, the empirical studies of Ludlow, Singapore and Nanjing have elaborated in which way, and to what extent, these two approaches could be integrated in investigating the historical development and socio-economic performance of urban landscape.

Secondly, an analytical method that combines and coordinates the historico-geographical and the configurational approach has been developed. On one hand, the integrated approach could contribute to a more precise plan unit recognition, in which the generative process of urban growth and development is constructed and analyzed in a specific urban context. This could improve the understanding of the historico-geographical structure of city as an interrelated entity in a process of change.

On the other hand, the integrated approach could contribute to a cross-scale landscape unit evaluation, in terms of historicity and spatial potential (how urban form supports, impedes or organizes socio-economic activities). This evaluation considers spatial potential apart from historic character, providing a balanced understanding of the fast-changing urban landscape. Meanwhile, in measuring spatial potential, the emerging open data (POIs) and functional mix method (Hoek, 2010) allow a more accurate and finer-scale description of how existing socio-economic activities are organized in the built environment. By linking the functional mix with the configurational analysis, the capacity of urban form to accommodate socio-economic activities are uncovered, and this information is valuable for future development. In sum, with the help of the emerging open data and new analytical method, the integrated approach could contribute to a finer-scale evaluation of the potential socio-economic performance of landscape units, rather than using conventional land use data to describe the existing condition in a coarse resolution.

Thirdly, this analytical method could provide reference for urban landscape management in a small-scale and incremental way. In the conventional planning practice, there is a lack of historico-geographical awareness among urban planners. Historic features are treated like isolated elements. How they relate to one another and form the entire settlement is unclear. The management strategy is often based on the building values. Some historic and heritage buildings are conserved while others are demolished, resulting in a fragmented urban landscape. Land use zoning is perceived as the prevalent top-down approach to improve socio-economic performance. Yet it separates different uses and prevents a diverse urban community. Massive areas created are devoted to single uses (Christopher, 1966; Scheer, 2010).

The findings of this study could help manage change based on a finer division of landscape units according to their historicity and spatial potential. The overlapped mappings of landscape unit and spatial potential produce clusters of physical plots with varied spatial potential in the lowest hierarchy of landscape units. Based on this mapping and classification, different functional and spatial interventions could be suggested to improve or maintain their spatial potentials to support vibrant urban life.

Moreover, in this study, physical plot is both the analytical unit and the management unit, which could provide opportunities for finer-scale and incremental interventions, and for plausible management proposals in urban planning and design practice. The findings may ultimately contribute to the dialectical discourse between conservation and development, thus maintaining the historic expressiveness of both local areas and city as a whole and producing forms of development that are both attractive and marketable.

9.3 Limitations

Current study only focuses on the combination of the historico-geographical approach and the configurational approach as a first step of combining the four morphological traditions, including the other two approaches: the process-typological approach and the spatial analytical approach.

Due to the data availability, only the plan unit division of Ludlow and Chinatown has been studied. The landscape unit division and evaluation of the two cities have not been carried out. The analytical framework of the integrated approach has not been as equally applied as in the case study of Nanjing. Meanwhile, only three cities (Ludlow, UK; Chinatown, Singapore and Nanjing, China) in the Western and Asian contexts are taken as empirical studies to explore in which way and to what extent the historico-geographical and configurational approach could be combined. Yet to draw sound conclusions require more case studies to test and enrich this analytical framework that combines and coordinates these two approaches.

Due to the lack of historical records and cartographic maps of the traditional residential settlement in the nineteenth and early twentieth century Nanjing, depictions of the built environment of Old Chengnan (street, plot and building block) are difficult. Reconstruction of the detailed changes to urban landscape remains at a coarse level in this study. Thus, the landscape unit division is limited to three-tiers, and evidence to further divide it into settlement clusters is lacking.

In the analytical framework, the spatial potential of urban form in this study is defined by two variables: spatial accessibility and spatial capacity. Spatial capacity measures the ability of urban space to carry certain contents within a certain walking distance. The contents as indicators of socio-economic performance could be the number and differentiation of people and things, such as plot number, floor area or building density. In this study, the content is limited to Hoek (2010)'s framework, which measures the floor areas of live/work/visit functions as the indicator of urban diversity. This may not represent the real urban diversity based on people's activities and movement.

Meanwhile, the extent to which the three types of spatial potential—matched, in-between and mismatched—could reflect the variations of socio-economic performance in urban areas has been verified by the correlation between spatial accessibility and the functional mix. This is built on the predictive power of spatial accessibility in identifying socio-economic vital areas, which has been proved in different cultural contexts and urban layouts in the world (Hillier, 1996). Yet urban population density as a proxy for the real socio-economic performance has not been explored in this study, which is partially attributed to the coarse resolution of the open data Baidu heatmap. This deficiency cannot reflect the different population density in the neighbourhood scale.

Lastly, apart from historicity and spatial potential of urban form, there are also various factors that can potentially influence management of change, such as zoning laws, planning regulations, socio-economic development strategies, public participation and property ownership. Much more needs to be done in the future to ensure successful management proposals and practice.

9.4 Future Research

This book is just a first step in a long journey towards combining and coordinating different morphological approaches to enhance the understanding, explanation and prescription of urban form. More rigorous and more empirical-based studies could be carried out to verify and enrich this analytical framework in the future. The dialectic between urban morphology and building typologies could be explored, including the relationship between urban areas and individual buildings, and the continuity and change of building forms. This enriched understanding may contribute to a finer delimitation of landscape units into several tiers.

Other factors such as building property right could be included and explored in this framework. It could help uncover how many households are included in one physical unit (courtyard), how they vary in building value and what their spatial potential typology is. This may provide a basis for private owners, government, planners and developers to participate in the decision-making process and negotiate on the management of change in a small-scale and incremental way.

The integrated framework developed in this book also has the potential to open up inquiries into social, economic and environmental phenomena with the emergence of new techniques and open data. For instance, the check-in data from social media, or GPS tracking-records of people's movement, provide a detailed description and measurement of people's real activities in urban space. Its correlation with the urban morphological structure may provide more powerful evidence to propose human-focused physical interventions and predict the outcomes of such interventions.

As Marshall (2009) argues, 'the most important role for urban design is not so much to tear the old down and build anew but to offer a different view of the future based on what is already happening, to develop alternative scenarios that are both plausible and exciting, to incite change through imagination'. Looking forward, there is the prospect of a stronger development arising out of the integration of different approaches, and urban morphology interpreted as a discipline has the potential to meet the challenge in the fast-changing urban landscape and to be embraced by a wider audience.

References

Baker, N. (2010). A characterisation of the historic townscape of Central Hereford. *Hereford: Herefordshire County Council. Herefordshire Archaeological Report, 266*.

Bandarin, F., & Van Oers, R. (2012). *The historic urban landscape: Managing heritage in an urban century*. Wiley.

Bobkova, E., Marcus, L., & Pont, B. (2017). *Multivariable measures of plot systems: Describing the potential link between urban diversity and spatial form based on*. Paper presented at the 11th Space Syntax Symposium, Lisbon.

Chen, F., & Thwaites, K. (2013). *Chinese urban design: The typomorphological approach*. Ashgate Pub. Ltd.

Christopher, A. (1966). A city is not a tree. *Design Magazine, 206*, 46–55.

Fang, Z., Hua, X., Liu, Q., Xu, Y., Du, L., & Chen, W. (2012). Jiucheng baohu de zaisikao [Reflections on the conservation of the historic city of Nanjing]. *Shijie Jianzhu Daobao, 2*, 10–33. (in Chinese).

Garyfalia, P. (2017). Urban morphology and World Heritage practice. *Urban Morphology, 21*(1), 83–86.

Griffiths, S., Jones, C. E., Vaughan, L., & Haklay, M. (2010). The persistence of suburban centres in Greater London: Combining Conzenian and space syntax approaches. *Urban Morphology, 14*(2), 85–99.

Gu, K. (2010). Urban morphological regions and urban landscape management: The case of central Auckland, New Zealand. URBAN DESIGN International, 15(3), 148–164.

Hillier, B. (1996). Cities as Movement Economies. Urban Design 1(1), 41–41.

Hillier, B., Turner, A., Yang, T., & Park, H. T. (2007). *Metric and topo-geometric properties of urban street networks*. Paper presented at the Proceedings Space Syntax. 6th International Space Syntax Symposium, Istanbul.

Hoek, J. V. D. (2010). The Mixed Use Index (MXI) as Planning Tool for (New) Towns in the 21st Century. In M. Provoost (Ed.), *New towns for the 21st century: The planned versus the unplanned city*. Martien de Vletter, SUN.

Jacobs, J. (1961). *The death and life of great American cities*. Wiley.

Karimi, K., Amir, A., & Shaiei, K. (2007). *Evidence-based spatial intervention for regeneration of informal settlements: The case of Jeddah central unplanned areas*. Paper presented at the 6th International Space Syntax Symposium, Istanbul.

Koh, M., & Melic, K. (2017). *A historic heart how heritage districts can make cities more liveable*. Retrieved from Singapore: https://www.clc.gov.sg/docs/default-source/reports/a-historic-heart.pdf

Kropf, K. (1996). An alternative approach to zoning in France: Typology, historical character and development control. *European Planning Studies, 4*(6), 717–737.

Kropf, K. (2009). Aspects of urban form. *Urban Morphology, 13*(2), 105–120.

Kropf, K. (2011). Morphological investigations: Cutting into the substance of urban form. *Built Environment, 37*(4), 393–408. https://doi.org/10.2148/benv.37.4.393

Kropf, K. (2014). Ambiguity in the definition of built form. *Urban Morphology, 18*(1), 41–57.

Larkham, P. J. (1996). *Conservation and the city*. Taylor & Francis.

Maffei, G., & Whitehand, J. (2001). Diffusing caniggian ideas. *Urban Morphology, 5*(1), 47–48.

Marcus, L. (2007). *Spatial capital and how to measure it: An outline of an analytical theory of the social performativity of urban form.* Paper presented at the Proceedings, 6th International Space Syntax Symposium, Istanbul.

Marcus, L. (2010). Spatial capital: A proposal for an extension of space syntax into a more general urban morphology. *The Journal of Space Syntax, 1*, 30–40.

Marshall, S. (2009). Cities design and evolution: Routledge London/New York.

Marcus, L., Pont, M. B., & Bobkova, E. (2017). *Cities as accessible densities and diversities.* Paper presented at the 11th Space Syntax Symposium.

Monteiro, C., & Pinho, P. (2021). MAP: A methodology for morphological analysis and prescription. *Urban Morphology, 25*(1), 57–75.

Moudon, A. V. (1997). Urban morphology as an emerging interdisciplinary field. *Urban Morphology, 1*(1), 3–10.

Oliveira, V., Monteiro, C., & Partanen, J. (2015). A comparative study of urban form. *Urban Morphology, 19*(1), 73–92.

Pinho, P., & Oliveira, V. (2009). Different approaches in the study of urban form. *Journal of Urbanism: International Research on Placemaking and Urban Sustainability, 2*(2), 103–125. https://doi.org/10.1080/17549170903083676

Scheer, B. (2010). *The evolution of urban form: Typology for planners and architects.* Routledge.

Scheer, B. C. (2016). The epistemology of urban morphology. *Urban Morphology, 20*(1), 5–17.

Shan, J. (2006). Cong daguimo jiufang gaizao dao xunxi jianjin youji gengxin [From mass demolish to incremental redevelopment]. *Wenwu* (7), 26–40. (in Chinese)

Stonor, T. (2017). A human-focused "science of building and cities" in the context of new techniques and new data. *Time+Architecture* (5), 63–66.

Thomas, R. M. (2018). Conservation, heritage and urban morphology-further thoughts. *Urban Morphology, 1*(22), 71–73.

Wang, L., & Mo, C. (2017). Chengshi gengxin he fengmao baohu de chengshi sheji yu chengshi zhili shijian [Urban Design And Governance For Urban Renewal And Landscape Preservation]. *Guihuashi, 33*(10), 135–141. (in Chinese).

Whitehand, J., & Gu, K. (2007a). Extending the compass of plan analysis: A Chinese exploration. *Urban Morphology, 11*(2), 91–109.

Whitehand, J., Gu, K., Whitehand, S. M., & Zhang, J. (2011). Urban morphology and conservation in China. *Cities, 28*(2), 171–185.

Whitehand, J. W. R. (2007). *Conzenian urban morphology and urban landscapes.* Paper presented at the Proceedings, 6th International Space Syntax Symposium, Istanbul.

Whitehand, J. W. R. (2015). Seeking an integrated approach to urban form: Tasks ahead. Urban Morphology, *19*(1), 3–4.

Whitehand, J. W. R., & Gu, K. (2007b). Urban conservation in China: Historical development, current practice and morphological approach. *Town Planning Review, 78*(5), 643–670.

Wu, L. (1994). *Beijing Jiucheng yu jier hutong [The old city of Beijing and Juer Hutong].* Beijing: Zhongguo jianzhu gongye chubanshe. (in Chinese)

Zhang, Y., Liu, Q., & Zhang, J. (2011). Lishi chengqu chiduxia de lishididuan xitong baohu zhengzhi gongzuofangfa tantao [The study of the conservation strategies of historic urban areas-based on the Conservation Planning of Old Chengnan]. *Chengshi guihua lunwenji*, 8295–8305 (in Chinese).

Index

X. Li and Y. Zhang, *Conserving and Managing Historical Urban Landscape*, The Urban Book Series, https://doi.org/10.1007/978-981-19-4222-8

GPSR Compliance

The European Union's (EU) General Product Safety Regulation (GPSR) is a set of rules that requires consumer products to be safe and our obligations to ensure this.

If you have any concerns about our products, you can contact us on ProductSafety@springernature.com

In case Publisher is established outside the EU, the EU authorized representative is:

Springer Nature Customer Service Center GmbH
Europaplatz 3
69115 Heidelberg, Germany

Batch number: 10726135

Printed by Printforce, the Netherlands